Bringing E-money to the Poor

DIRECTIONS IN DEVELOPMENT
Finance

Bringing E-money to the Poor

Successes and Failures

Thyra A. Riley and Anoma Kulathunga

WORLD BANK GROUP

1 2 3 4 20 19 18 17

ISBN (paper): 978-1-4648-0462-5
ISBN (electronic): 978-1-4648-1089-3
DOI: 10.1596/978-1-4648-0462-5

Cover photo: © BRAC. Used with the permission of BRAC; further permission required for reuse.
Cover design: Debra Naylor, Naylor Design, Inc.

Library of Congress Cataloging-in-Publication Data has been requested.

Contents

	Foreword	*xi*
	Acknowledgments	*xiii*
	About the Authors	*xv*
	Abbreviations	*xvii*
	Overview	**1**
	Background	1
	Motivation and Evidence	3
	Target Audience	5
	Methodology: Country Selection and Financial Inclusion Status	5
	Organization of This Volume	10
	Notes	12
	Bibliography	13
PART I	**Journey toward a Cash-Lite Society and Financial Inclusion**	**15**
Chapter 1	**The Challenge of Financial Inclusion**	**17**
	What Is Financial Inclusion?	17
	Why Does Financial Inclusion Matter?	20
	The Global Financial Inclusion Gap	22
	South Asia's Financial Inclusion Gap	23
	Poverty, Financial Exclusion, and Financial Vulnerability in South Asia	27
	Remittance Transfers and Financial Inclusion	29
	Notes	32
	Bibliography	33
Chapter 2	**Digitizing Financial Inclusion through Innovations**	**37**
	Types of Innovation for Financial Inclusion	37
	E-money and Digital Payments	40
	Toward a Cash-Lite Society	42

	Risks in Digital Finance	46
	Notes	51
	Bibliography	51
Chapter 3	**Stakeholders in Digital Financial Inclusion**	**53**
	Introduction	53
	Macro-Level Stakeholders: Policy Makers, Regulators, and Donors	54
	Meso-Level Stakeholders: Enabling Institutions	55
	Micro-Level Stakeholders: Institutions Offering Digital Solutions	56
	Customer-Level Stakeholders: Users	57
	Bibliography	57
PART II	**Critical Enablers That Are Game Changers in Successful E-money Deployments**	**59**
Chapter 4	**Policy Leadership and Enabling Regulatory Environments**	**63**
	Introduction	63
	Regulatory Balance in Financial Innovation	64
	Kenya: Leadership Lesson from the Central Bank of Kenya	65
	India: Jan Dhan Yojana Flagship Financial Inclusion Plan	69
	Sri Lanka: Regulations Keeping Pace with Technological Advancements	76
	Thailand: A Government's Vision and Policy to Bring Cash to the Doorstep	84
	The Philippines: The World's Oldest Mobile Money Initiative Has Yet to Reach Potential	91
	Maldives: Mobile Money Opportunity Still Knocking at the Door	94
	Notes	98
	Bibliography	101
Chapter 5	**Innovative Uses of Infrastructure and Digital Ecosystems**	**105**
	Introduction	105
	Interoperability in Indonesia, Pakistan, Sri Lanka, Tanzania, and Thailand	105
	Agent Network Management in Kenya	117
	Digitizing Social Grant Disbursement Programs: Brazil, Mexico, and South Africa	141
	Notes	150
	Bibliography	151

Chapter 6	**Unique Identification**	**155**
	Introduction	155
	The Philippines: 21 IDs and Counting	158
	India's Aadhaar Program: Potential Game Changer in Digital Financial Inclusion	163
	Sri Lanka: Mobile Connect, the Interoperable ID	171
	Notes	174
	Bibliography	174
PART III	**South Asia Digital Landscape, Future Options, and Conclusions**	**177**
Chapter 7	**Digital Landscape in South Asia**	**179**
	Introduction	179
	Macro-Level Strategies	181
	Meso-Level Approaches and Issues	182
	Micro-Level Models	183
	Customer-Level ID Systems	184
	Annex 7A Digital Financial Landscape in South Asia, by Country: At a Glance	186
	Note	193
	Bibliography	193
Chapter 8	**Opportunities, Challenges, and Future Options in South Asia**	**195**
	Introduction	195
	Macro Level	195
	Meso Level	197
	Micro Level	197
	Customer Level	199
	Note	199
	Bibliography	199
Chapter 9	**Conclusions**	**201**
	Introduction	201
	Role of Governments and Regulators	202
	Coordinated Action, Common Platforms, and Interoperability	205
	Outreach by Retail Institutions	206
	Increasing Accessibility for Customers	207
	The Journey toward a Cash-Lite Society: Coordination and Balance	208
Appendix A	**Findex Data for Selected Countries**	**211**

Boxes

1.1 "Financial Inclusion": A Working Definition 18
2.1 Cash versus Electronic Payments 43
2.2 Doing Digital Finance Right: The Case for Stronger Customer Risk Mitigation 50
4.1 M-Pesa: A Backstory and an Alternative Perspective 67
4.2 Reserve Bank of India Regulatory Reforms, 2014 76

Figures

1.1 Share of Adults with a Financial Services Account, by Region, 2014 24
1.2 Share of South Asian Adults with a Financial Services Account, by Country, 2014 25
1.3 Share of South Asian Adults with a Financial Services Account, by Gender and Country, 2014 26
1.4 Access to Finance in South Asia: Supply-Side Data, 2010 27
1.5 Poverty, Financial Exclusion, and Financial Vulnerability Indicators in South Asia, 2014 28
1.6 Remittances and Other Resource Flows to Developing Countries, 1990–2015 30
2.1 Sample Relative Costs of Payment System Infrastructure, from Bank Branches to Mobile Phone 43
2.2 Stages and Shifts from a Cash-Heavy to a Cash-Lite Society 45
II.1 Number of Active Mobile Money Services Worldwide, by Region, 2001–14 61
4.1 Financial Access Strand in Kenya, 2006 65
4.2 Financial Access Trends in Kenya, 2006–13 68
4.3 Use of Financial Services in Kenya, by Type, 2006–13 69
4.4 Zero-Balance Trends in Jan Dhan Yojana Accounts, India, 2014–15 73
4.5 Number of 2G and 3G/4G Connections in India, 2008–17 75
4.6 Financial Access Strand in Thailand, 2013 85
4.7 Financial Access Strand in Thailand, by Region, 2013 85
4.8 Average Time to Financial Service Touchpoints in Thailand, 2013 89
5.1 Market Share of Sri Lankan Mobile Service Providers, 2014 109
5.2 Schematic of End-to-End Interoperable eZ Cash System 111
5.3 Comparing Mobile Money Use in Tanzania and Kenya, 2007–13 113
5.4 Active Subscriber Market Shares of Tanzanian Mobile Service Providers, 2014 114
5.5 Financial Account and Mobile-Phone Penetration, Indonesia versus Selected Asian Countries, 2014 115
5.6 Mobile Money Awareness in Indonesia, 2014 116

5.7 Number of Financial Access Points across Developing Countries, 2014 118
5.8 Growth in Number of M-Pesa Customers and Agents, 2007–14 120
5.9 Initial M-Pesa Agent Network Structure 123
5.10 M-Pesa Agent Network Structure with Formal Introduction of Aggregators 124
5.11 Current M-Pesa Agent Network Structure and E-float/Cash Management Process 126
5.12 Mobile Money Transfer Value Chain 129
5.13 M-Pesa Service Development, 2007–13 130
5.14 Growth in Number of M-Shwari Savings Accounts, 2013–14 136
5.15 Average Capital Expenditure Costs for Financial Service Providers in Kenya, by Channel 140
5.16 Share of Adults Receiving Government Transfers, by Region and Payment Method, 2014 142
5.17 Financial Access Strand in South Africa, 2004–14 147
6.1 Use and Awareness of Payment System Providers in the Philippines, 2010 162
6.2 Aadhaar Registration Trends in India, 2014–15 164
6.3 Top 10 States for Aadhaar Registration in India, 2015 164
6.4 Aadhaar Registration, by Gender and Age Group in India, 2015 165
6.5 Number of Aadhaar Registrations Completed by Top 10 Service Providers in India, May 2015 166
6.6 Financial Inclusion Applications of Aadhaar 166
6.7 Mobile Connect Beta Trial Indicators 173

Maps

O.1 Universal Financial Access 2020 Focus Countries 2
4.1 Distribution of Financial Institution Branches, Automated Machines, and EFTPOS Terminals in Thailand, by Region, 2013 87
5.1 Number of Live Mobile Money Services for the Unbanked, by Country, 2014 128
6.1 Global Participation in Biometric ID Programs, by Region, 2012 157

Tables

O.1 Selection Criteria for Case Study Countries 6
O.2 Use of Transaction Accounts, Case Country Comparison, 2014 7
1.1 Estimated Financial Inclusion Gap, Globally and by Region, 2008 23
1.2 South Asia Remittance Receipts, by Country, 2009–13 31
2.1 Differences between Electronic Money and Virtual Currency Schemes 41
4.1 Jan Dhan Yojana Account Status, by Bank Type, May 2015 71

4.2	Financial Service Providers in Thailand, by Customer and Transaction Type, 2013	90
4.3	Cash in Circulation in Maldives, 2007–14	95
5.1	Financial Inclusion in Sri Lanka Relative to South Asia and Lower-Middle-Income Countries, 2014	108
5.2	Transaction Cost Comparison for eZ Cash and mCash	109
5.3	Roles and Responsibilities in M-Pesa Agent Structure	120
5.4	Detailed Roles and Responsibilities in M-Pesa Agent Structure	121
5.5	Timeline of M-Pesa Expansion of Functionality and Services, 2005–13	131
5.6	Banks and MFIs Linked to M-Pesa, 2013	133
5.7	Key M-Shwari Statistics	136
5.8	Payment Approaches of Selected Grant Programs as of 2012	143
5.9	Characteristics of the SASSA Card	148
6.1	Acceptable ID Documentation for Financial Services in the Philippines	160
6.2	Key Differences between Aadhaar and the National Population Register	170
7.1	Financial Inclusion Data by Region, 2014	180
7A.1	Digital Financial Landscape in South Asia: At a Glance	186
A.1	India against Benchmarks for South Asia and Lower-Middle-Income Countries	211
A.2	Indonesia against Benchmarks for East Asia and Pacific and Lower-Middle-Income Countries	212
A.3	Kenya against Benchmarks for Sub-Saharan Africa and Low-Income Countries	214
A.4	The Philippines against Benchmarks for East Asia and Pacific and Lower-Middle-Income Countries	215
A.5	South Africa against Benchmarks for Sub-Saharan Africa and Upper-Middle-Income Countries	216
A.6	Sri Lanka against Benchmarks for South Asia and Lower-Middle-Income Countries	218
A.7	Thailand against Benchmarks for East Asia and Pacific and Upper-Middle-Income Countries	219

Foreword

Financial inclusion can almost be taken as a given in high-income countries, but access to finance often remains sporadic and informal in low- and middle-income countries. And yet, there is clear evidence that financial inclusion accelerates economic growth and enhances opportunities, especially among the poor. Communities thrive when households and local businesses gain access to financial services.

In recognition of these benefits, the World Bank and the International Monetary Fund launched in 2015 the Universal Financial Access 2020 (UFA 2020) initiative. UFA2020 aims to enable 2 billion financially excluded adults to gain access to transaction accounts. The initiative focuses on 25 countries where 73 percent of all the financially excluded people live.

Three South Asian countries—Bangladesh, India, and Pakistan—account for 30 percent of the world's financially excluded population and represent 40 percent of the UFA2020 target population. Thus, the South Asia region draws particular attention when it comes to broadening access to finance.

India's transformational efforts in implementing the Aadhaar financial inclusion program and the unique identification program show that success is possible. Thanks to new technologies, transformative business models, and ambitious reforms, universal access to financial services has evolved from an aspirational goal to a target within reach.

Improvements in the legal, regulatory, and institutional environments—which tend to be useful for development in general—can have a favorable effect on financial inclusion. Also, policy makers can promote financial inclusion by supporting innovative business models that increase the outreach and lower the cost of payment and financial services.

Bringing E-money to the Poor: Successes and Failures reviews the experiences of countries that have demonstrated notable success in applying new technologies and institutional innovation to provide the poor and vulnerable with entry points into the financial system. Its case studies are based on extensive field research and interviews with financial sector practitioners, users, policy makers, and regulators. Detailed contextual analysis and an emphasis on critical conditions help identify the drivers of success, as well as the challenges and risks.

Although new technologies and innovative methodologies in the finance industry are numerous, the study focuses on e-money initiatives such as mobile money, interoperable and multifunctional automated teller machines (ATMs), and prepaid debit cards for social grant programs.

The four cases selected—Kenya, South Africa, Sri Lanka, and Thailand—illustrate the importance of leadership by the authorities, innovation by the private sector, and a flexible "learn before regulating" approach. The result has been a transformative expansion of financial access not only to the poor but also throughout the economy, as these case studies show:

- In *Kenya*, the rate of financial inclusion more than doubled in five years to reach nearly 70–75 percent of the adult population. Innovation took the form of a mobile money application (M-Pesa) by the country's leading mobile phone operator. A cooperative and enabling relationship between the regulator and the operator helped M-Pesa become the country's retail payment platform.

- In *South Africa*, the key innovation was the use of a biometrically secure, "chipped" open-debit MasterCard as the platform for social transfer payments. Financial access was extended to 10 million of the country's poor, pushing financial inclusion up to 86 percent of the population. The essential element of success was the cooperation between the Social Security Agency, the private creator of the biometrically enabled card, and a local bank.

- In *Sri Lanka*, a proactive development of the legislative framework enabled the establishment of an excellent payment systems infrastructure. Sri Lanka arguably has the best regulatory framework in the South Asia region to govern e-money for e-commerce and e-government, as well as the world's first end-to-end interoperable payment solution. A range of private sector players and mobile operators jumped in, and financial inclusion is already reaching over 83 percent of the population.

- In *Thailand*, 88 percent financial inclusion of households has been achieved through efficient coordination of strategies and policies toward payment services and reduction of infrastructure costs. Thousands of multicapacity ATMs and automated deposit machines (ADMs) were deployed throughout the country as a result. The leadership of the Thai Bankers' Association was a key element of this success.

We hope that these rich case studies stimulate debate and encourage policy makers, regulators, financial service providers, and mobile network operators to move forward on access to finance, especially for the poor. Their initiative, enthusiasm, and cooperation are needed to make universal financial inclusion a reality in South Asia.

Martin Rama
Chief Economist, South Asia Region
The World Bank

Acknowledgments

This study was written by Thyra A. Riley (now retired) in her role as sector coordinator and lead specialist, and by Anoma Kulathunga, senior financial sector specialist—both of the World Bank's Finance and Markets Global Practice, South Asia region.

The authors are especially grateful to Martin Rama, chief economist of the South Asia region, who provided invaluable support and guidance as the chairman of the peer-review process by which this program of study was undertaken and published. The authors also thank Henry Bagazonzya and Niraj Verma, practice managers of the South Asia region's Finance and Markets Global Practice, under whose supportive auspices this product was brought to final fruition.

The authors express special appreciation for the ex officio moral support and intellectual contributions of Christopher Dooley (senior adviser, United Nations Capital Development Fund); Ranee Jayamaha (former deputy governor of the Central Bank of Sri Lanka); William F. Steel (adjunct professor, University of Ghana, Legon, and former World Bank senior adviser on microfinance and small enterprise); and Martin Melecky (lead economist, Office of the Chief Economist, South Asia region).

Valuable feedback and comments were provided by World Bank peer reviewers: Simon Bell, global lead, SME Finance, Finance and Markets Global Practice; Martin Kanz, economist, Development Economics Chief Economist's Office; Harish Natarajan, lead financial sector specialist, Payment Systems Development Group, Finance and Markets Global Practice; Douglas Pearce, practice manager, Financial Infrastructure and Access, Finance and Markets Global Practice; and Maja Andjelkovic, senior financial sector specialist on behalf of the Innovation and Entrepreneurship Team of the Trade and Competitiveness Global Practice.

Country e-money landscapes were prepared by World Bank or International Finance Corporation Country Office colleagues: Sabin Shrestha, senior financial sector specialist; Nazir Ahmad III, private sector specialist (Afghanistan); and Santosh Pandey, senior financial sector specialist (Nepal); as well as by country-based financial consultants: Muhymin Chowdhury (Bangladesh); Ranee Jayamaha (Maldives); Caroline Pulver (Kenya); and K. R. Ramamoorthy (India).

The authors are deeply indebted to the World Bank publications team: Jewel McFadden, acquisitions editor; Paola Scalabrin, acquisitions editor; Stephen Pazdan, publishing associate; Mary A. Anderson, our copy editor; and Gwenda Larsen, our proofreader.

Last, and most important, the authors recognize the invaluable contributions and vision of the business leaders, central bank regulators, and international donors that incorporate e-money as a delivery mechanism to provide access to financial services to the poor. The cases and frameworks discussed in this study are built on the authors' in-country fieldwork and interviews with these leaders, their staff, their clients, and the users of e-money. The generous access provided has made the richness of this study possible—with our very sincerest thanks!

About the Authors

Thyra A. Riley, now retired, was sector coordinator and lead specialist of the World Bank's Finance and Markets Global Practice, South Asia region. Over a 30-year career at the World Bank, she served in several corporate management, knowledge management, and lead financial sector specialist positions in several regions of the world, including Africa, Latin America and the Caribbean, and Middle East and North Africa. As the knowledge manager for the Micro, Small Enterprise, and Rural Finance Thematic Group, she led knowledge-sharing engagements that brought together leading international microfinance practitioners with African country leaders, policy makers, and donors interested in learning from the best-practitioners themselves. Riley also led several projects and knowledge-sharing engagements with the South African government during the postapartheid development of the country's policy framework for micro and small enterprises. She has written extensively about lessons learned from high-impact development interventions, focusing on approaches that have mainstreamed access by the poor to financial services through innovative means including traditional microfinance and digitally enabled financial services. She was a visiting fellow in finance at the Sloan School of Management, Massachusetts Institute of Technology. Riley holds a bachelor's degree in development economics from Stanford University and a master's degree in public and international affairs from Princeton University.

Anoma Kulathunga is a senior financial sector specialist in the World Bank's Finance and Markets Global Practice, South Asia region. During her 12 years at the World Bank, she brought her financial sector expertise to numerous projects, including country experience spanning all South Asian countries, the Middle East and North Africa, Indonesia, Uganda, and Vietnam. Before joining the World Bank, she served for 11 years as regulator at the Central Bank of Sri Lanka and has also been an assistant professor of finance at The George Washington University, Washington, DC. An associate member of the Chartered Institute of Management Accountants, UK, she has coauthored five books and published many papers on issues related to financial stability and soundness. Her research interests include financial sector development and stability, financial infrastructure, Islamic banking, worker remittances, and international banking. Kulathunga holds an MBA from the University of Sri Jayewardenepura, Sri Lanka, and master's and doctoral degrees in international finance and development economics from The George Washington University.

Abbreviations

ACH	automated clearinghouse
ADM	automated deposit machine
AEPS	Aadhaar Enabled Payment System
AML/CFT	Anti-Money Laundering and Combatting Funding of Terrorism
ANM	agent network manager
API	application programming interface
ASEAN	Association of Southeast Asian Nations
ATM	automated teller machine
BC	business correspondent
BDO	Banco De Oro (Philippines)
BFP	Bolsa Família Program (Brazil)
BI	Bank of Indonesia
BIS/CPMI	Bank for International Settlements Committee on Payments and Market Infrastructures
BML	Bank of Maldives
BOT	Bank of Thailand
BSP	Central Bank of the Philippines (Bangko Sentral ng Pilipinas)
B2P	business-to-person
CBA	Commercial Bank of Africa
CBK	Central Bank of Kenya
CBSL	Central Bank of Sri Lanka
CCAPS	Common Card and Payment Switch
CDD	customer due diligence
CDM	cash deposit machine
CDMA	Code Division Multiple Access
CEB	Ceylon Electricity Board
CGAP	Consultative Group to Assist the Poor
CI/CO	cash-in/cash-out
CITSG	Core Information Technology Support Group (Philippines)

CNIC	Computerized National Identity Card
COSO	Committee of Sponsoring Organizations of the Treadway Commission
CPMI	Committee on Payments and Market Infrastructures (of the Bank for International Settlements)
CPS	Cash Paymaster Services
CSIRT	Computer Security Incident Response Team
CSP	certification service provider
DBT	direct benefits transfer
DFID	Department for International Development (United Kingdom)
DFS	digital financial service
EFT	electronic funds transfer
EFTPOS	electronic funds transfer at point of sale
EMI	e-money issuer
EMV	Europay, MasterCard, and Visa
FATF	Financial Action Task Force
FDI	foreign direct investment
FMI	financial market infrastructure
FMIS	financial management information system
FSD	Financial Sector Deepening
GCC	Gulf Cooperation Council
GDP	gross domestic product
GNI	gross national income
GSM	Global System for Mobiles
GSMA	Groupe Speciale Mobile Association
G2C	government-to-consumer
G2P	government-to-person
G20	Group of Twenty (countries)
IBFT	Inter Bank Fund Transfer
ICT	information and communication technology
ICTA	Information and Communication Technology Agency (Sri Lanka)
ID	identification
IFC	International Finance Corporation (of the World Bank Group)
IMF	International Monetary Fund
IMPS	Immediate Payment Service (India)
IOM	International Organization for Migration
IT	information technology
ITMX	Interbank Transaction Management and Exchange (Thailand)
JDY	Jan Dhan Yojana (India)

KCB	Kenya Commercial Bank
KYC	know your customer
LECO	Lanka Electricity Company
LPG	liquefied petroleum gas
M-Pesa	Kenya's mobile money platform
M-POS	mobile point of sale
MFI	microfinance institution
MGNREGA	Mahatma Gandhi National Rural Employment Guarantee Act
MMA	Maldives Monetary Authority
MMU	Mobile Money for the Unbanked
MNO	mobile network operator
MoRD	Ministry of Rural Development (India)
MOU	memorandum of understanding
MPS	mobile payment system
MUDRA	Micro Units Development and Refinance Agency (India)
NADRA	National Database and Registration Authority (Pakistan)
NBFC	nonbanking finance company
NDB	National Development Bank (Sri Lanka)
NFC	near-field communication
NIC	national identity card
NPC	national payment council
NPCI	National Payments Corporation of India
NPR	National Population Register (India)
NSSLA	nonstock savings and loan association
NSSO	National Sample Survey Office (India)
NUUP	National Unified USSD Platform (India)
NWSDB	National Water Supply and Drainage Board (Sri Lanka)
ODA	official development assistance
OECD	Organisation for Economic Co-operation and Development
OTC	over-the-counter
PIN	personal identification number
POS	point of sale
PPP	purchasing power parity
PSSA	Payment and Settlement Systems Act (Sri Lanka)
P2B	person-to-business
P2G	person-to-government
P2P	person-to-person
PUM	passbook update machine
RBI	Reserve Bank of India

ROSCA	rotating savings and credit association
RTGS	real-time gross settlement
SAARC	South Asian Association for Regional Cooperation
SACCO	savings and credit cooperative
SASSA	South African Social Security Agency
SFI	specialized financial institution
SIM	subscriber identification module
SME	small and medium enterprise
SMS	short message service
TBA	Thai Bankers' Association
TRAI	Telecom Regulatory Authority of India
TRCSL	Telecommunications Regulatory Commission of Sri Lanka
UEPS	Universal Electronic Payment System
UFA2020	Universal Financial Access 2020
UID	unique ID
UIDAI	Unique Identification Authority of India
UMID	unified multipurpose ID
UNHCR	United Nations High Commissioner for Refugees
USSD	unstructured supplementary service data

Currencies

Currency conversions that appear in the text were current as of the time of writing (in 2015).

B	Thai baht
K Sh	Kenya shilling
R$	Brazilian real
R	South African rand
Rf	Maldivian rufiyaa
Rs	Indian rupees
SL Rs	Sri Lanka rupees
T Sh	Tanzania shilling

Overview

I'm confident digital tools will allow us to ignite significant progress, opening a broader path to the goal of universal access to basic accounts by 2020. But access is just an interim step. The full benefit of financial inclusion depends on households and small businesses actively using a range of affordable and effective financial services, coupled with financial education and consumer protection. That is a much taller order.

—Queen Máxima of the Netherlands, United Nations Secretary-General's Special Advocate for Inclusive Finance for Development, UNSGSA Annual Report 2015

Background

Financial sector development can aid economic growth and create private and social benefits, especially in countries at lower levels of development (Cull, Ehrbeck, and Holle 2014; Sahay et al. 2015). An important aspect of financial development to address the shared prosperity objective is the extension of financial intermediation services to low-income brackets of the population.

Access to financial and payment services, including savings, credit, and social welfare transfers, facilitates better financial inclusion and enables improved income distribution and inclusive growth. Although financial inclusion is mostly a foregone conclusion in the developed world, in developing countries it often remains sporadic or at best informal for those at the base of the pyramid.[1] Limited financial inclusion severely impacts financial stability, financial security, and poor people's economic mobility, thus effectively impeding the achievement of shared prosperity and development. The *Global Financial Development Report 2014* suggests that public policy can achieve potentially large effects on financial inclusion through reforms (World Bank 2014). The evidence provided from the World Bank's *Doing Business* data (World Bank 2017) and Enterprise Surveys[2] indicates that improvements in the legal, regulatory, and institutional environments—which tend to be useful for development in general—can also have a favorable effect on financial inclusion.

The Universal Financial Access 2020 (UFA2020) goal to enable 2 billion financially excluded adults to gain access to a transaction account—an initiative established during the 2015 World Bank and International Monetary Fund (IMF) Spring Meetings—focuses on 25 countries where 73 percent of all financially excluded people live (map O.1).[3] The largest shares of unbanked

Map O.1 Universal Financial Access 2020 Focus Countries

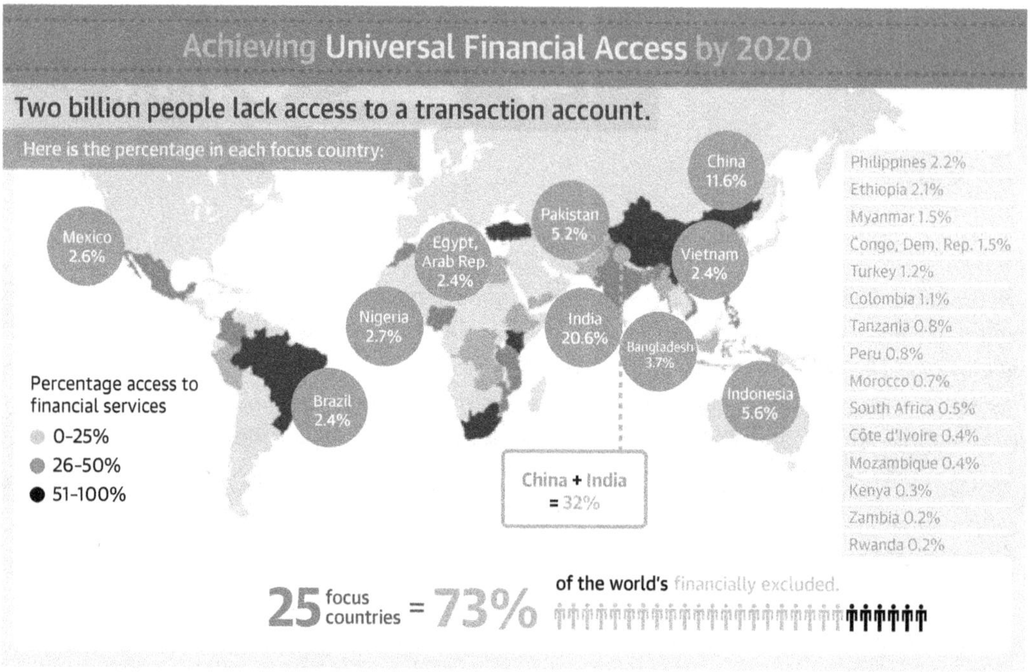

Source: World Bank, from 2014 Global Findex and International Monetary Fund (IMF) Financial Access Survey data. © World Bank. http://www.worldbank.org/en/topic/financialinclusion/brief/achieving-universal-financial-access-by-2020. Permission required for reuse.

people are in India (which accounts for about 21 percent of the world's financially excluded working-age population) and China (with about 12 percent). The other top-priority countries include Bangladesh and Pakistan, with about 4 percent and 5 percent, respectively, of the world's financially excluded population. Thus, three South Asian countries that account for 30 percent of the world's financially excluded people represent 40 percent of the UFA2020 target population.

South Asia plays a key role in the global development arena, with the world's largest working-age population, a quarter of the world's middle-class consumers, the world's greatest number of poor and undernourished people, and several fragile states of global geopolitical importance. Led by India, strong inclusive growth in South Asia could potentially change the face of global poverty. Although the modern microfinance industry—which emerged in South Asia in the 1970s with organizations such as Grameen Bank of Bangladesh—has contributed meaningfully to expanding outreach and access in the region, data show that the number of people with access to formal financial services falls short of the potential that we associate with the impressive levels of financial access and inclusive growth in the emerging markets of East Asia.

Demirgüç-Kunt, Beck, and Honohan (2008) show evidence that financial development and improved access to finance are likely not only to accelerate economic growth but also to reduce income inequality and poverty, and they describe how poor communities thrive economically when provided with access to financial services.

However, having more than 40 percent of the world's financially excluded people, it is clear that in South Asia traditional banking has failed to adequately reach the poor and the financially vulnerable. On the other hand, technological innovations have responded to changing consumer behavior and tightened bank regulations by offering alternative means to achieve inclusive finance, with some clear success stories.

This study is based on case studies—developed through in-depth field visits and desk research—that analyze the implementation of specific e-money and other digital payment programs. The primary criterion for selection was the availability of relevant data to serve as evidence in analyzing the country's experience. Countries were then selected that had demonstrated successful outcomes where critical enablers of success could be identified. In addition, countries were included that showed early promise but where critical constraints could be identified that stalled progress.

Motivation and Evidence

This study aims to identify countries that have demonstrated notable success in applying new e-money technologies and innovative thinking in providing *first entry points* into the financial system for poor and vulnerable population segments. Case studies are used to emphasize detailed contextual analysis of certain critical conditions and their relationships to the success or failure of these interventions. Although new technologies and innovative methodologies in the finance industry are numerous, the study narrowly focuses on e-money initiatives such as mobile money, interoperable and multifunctional automated teller machines (ATMs), and prepaid debit cards for social grant programs as the first entry points to financial inclusion.

The focus is on analyzing the provision of cost-effective, reliable, and safe access to basic cash-in/cash-out, utility, and bill payment services to financially unserved or underserved people through the selected e-money interventions. Although the study examines cases where financial intermediation activities such as credit, savings, insurance, and other financial products are developed through the e-money platforms, it does not cover the entire spectrum of financial inclusion at these entry points.

The observed outcomes from four of the selected successful country case studies—Kenya, South Africa, Sri Lanka, and Thailand—show that the private sector and nonbank entities have been supported or, in some cases, been led by flexibly designed policies and regulations, as in the following cases:

- *In Kenya,* the rate of financial inclusion more than doubled in five years to reach nearly 70–75 percent (depending on methodology) of the adult population as a direct result of innovations associated with a mobile money application (M-Pesa) that has evolved into the country's retail payment platform (FSD Kenya and CBK 2013).[4] While also used ubiquitously by the banked and well-to-do, M-Pesa has especially benefited the poor and unbanked who previously had limited and costly access to traditional bank and financial infrastructure.

- *In South Africa,* the use of a biometrically secure, "chipped" open-debit MasterCard as the platform for social transfer (government-to-persons [G2P]) payments extends financial access to 10 million of the country's poor.[5] This is the main contributing factor to the growth in the country's banked population from 63 percent in 2011 to 75 percent in 2014, with financial inclusion (adding those who use nonbank accounts for financial transactions) reaching 86 percent (FinMark Trust 2014).
- *In Sri Lanka,* the government and the Central Bank of Sri Lanka have proactively developed the country's legislative framework, enabling the establishment of an excellent payment systems infrastructure and possibly the best regulatory framework in the region to govern e-money for e-commerce and e-government. This policy approach has facilitated the launch of the world's first end-to-end interoperable mobile payment solution as another means of enhancing financial inclusion that is already reaching over 83 percent of the population.[6]
- *In Thailand,* 88 percent financial inclusion of households has been achieved through efficient coordination of strategies and policies toward payment services and reduction of infrastructure costs, partly through the deployment of thousands of multicapacity ATMs and automated deposit machines (ADMs) throughout the country (BOT 2014).[7]

This study also draws lessons from experiences in several other countries: India, Indonesia, Maldives, and the Philippines. Some of these countries have taken important initial steps to create the potential for rapid expansion of financial inclusion, while others have encountered obstacles that have limited their success.

In most instances, reform packages are country-specific. For this reason, there is considerable uncertainty as to which countries or initiatives reflect best practice. Nevertheless, pursuing cross-country studies of successful practices or policy initiatives, along with international dialogue, can flatten the learning curve and speed up policy learning by highlighting common traits, implementation issues, and operational successes. Furthermore, to effectively counter underlying barriers to financial access for underserved groups, even within a single country, it appears to be important to follow an integrated approach that considers the entire ecosystem at different stakeholder levels, as explained in chapter 3.

This study also aims to identify new approaches to improving financial inclusion in South Asia. It documents innovative uses of technology in the form of digital financial services operating within a balanced regulatory environment that can be key to improving financial inclusion. When an economy is cash-based, access to financial services is often restricted and costly. When countries move increasingly away from cash to a "cash-lite" economy—one in which "cash is no longer the most common means of payment" (BFA 2015)—broader financial inclusion can take place at a faster pace and lower cost. This is because a digital solution can often be more easily deployed to successfully address inclusion barriers and provide benefits in terms of proximity, safety, reliability, cost, and simplicity. These are hugely important concepts, especially for poor people. Technology or regulatory balance alone will not make an initiative successful unless it accrues one or more of these benefits to the users.

It is important to understand that although these initiatives represent exciting use of digital means in providing entry points to formal financial systems, they are but the first steps toward embarking on a journey toward a cash-lite economy. It is not an easy feat to develop inclusive digital financial systems and have a meaningful distribution effect among poor people, let alone reach the cash-lite or cashless stages that the developed countries have achieved. Although the viability of adopting digital solutions varies from market to market, these initiatives present a powerful opportunity to draw lessons to advance financial inclusion through efficient, affordable digital means.

Target Audience

The target audience for this study comprises national regulators, policy makers, and market practitioners of digital financial services, primarily in the South Asia region. Given the growing interest in e-money and digital payment solutions—and in light of ongoing initiatives in most South Asian countries—developing a knowledge-sharing platform both in the form of this volume and in a seminar would allow for an interactive knowledge exchange between practitioners from the case-study countries and their South Asian counterparts. The case study analysis, combined with descriptions of South Asian country financial and e-money landscapes, provides an excellent base for more-comprehensive diagnostic studies in the future on the digital financial potential of South Asian countries. Researchers, experts, and private sector service providers should also find the study informative.

Methodology: Country Selection and Financial Inclusion Status

Hawkins (1980) described an outlier as an observation that "deviates so much from other observations as to arouse suspicions that it was generated by a different mechanism." The selection of countries for the case studies was first triggered by the observance in the data of obvious outliers, as found in the Findex data,[8] combined with field visits and readings that confirmed that the chosen countries have indeed acted in different and innovative ways to achieve higher levels of

Table O.1 Selection Criteria for Case Study Countries

Case study country	*Evidence-based data*	*Demonstrated successful outcomes*	*Critical enabler of success identified*	*Critical constraint on progress identified*
Countries with documented success or constraints				
Kenya	X	X	X	
Maldives	X			X
Philippines	X			X
South Africa	X	X	X	
Sri Lanka	X	X	X	
Thailand	X	X	X	
Additional countries with promising early efforts				
India	X		X	
Indonesia	X		X	

basic financial access for their poor populations by consciously developing the financial and payment ecosystem using digital means.

This selection process yielded six suitable countries (table O.1): four fast movers (Kenya, South Africa, Sri Lanka, and Thailand) that have successfully pursued e-money initiatives that have also yielded transformational levels of financial inclusion, including use of e-money by the poorest;[9] and two early movers (Maldives and the Philippines) where the momentum implementing e-money has stalled or failed to reach its full potential but remains promising. In addition, the study reviews the possible game-changing initiatives undertaken by a South Asian country and a Southeast Asian country where critical enablers could be identified but the outcomes cannot yet be documented: specifically, India's establishment of a unique identification system and Indonesia's steps to achieve interoperability among providers to help overcome geographic challenges to outreach.

These case studies offer in-depth analysis of the elements of some innovative e-money initiatives that have influenced project outcomes and digital financial landscapes in terms of expansion of financial access and inclusion. The analysis focuses on stakeholders at all levels, highlighting common critical elements or "game changers" that fast movers have implemented, but which the early movers overlooked, preventing them from reaching their full potential.

Summary of Findex Data

The 2014 Global Findex Survey data on each of these countries (see appendix A) are used in the ex post outcome analysis. Table O.2 provides a comparative summary of country data on transactions (both inflow and outflow) that measure the use of, and hence the demand for, such accounts. These data points indicate the relationship between digital services and extension of payment services to underserved or poor populations and, also, shed light on the readiness of such countries and population segments to use digital technologies.

Table O.2 Use of Transaction Accounts, Case Country Comparison, 2014

Percentage of respondents ages 15 years and older

Survey question	*India*	*Kenya*	*Philippines*	*South Africa*	*Sri Lanka*	*Thailand*	*World average*
Paid school fees							
In the past year	22.4	51.3	40.5	26.6	29.8	26.4	—
In the past year, rural	20.5	52.8	41.1	23.7	28.3	23.2	—
In the past year, income, poorest 40%	21.7	50.6	44.8	21.7	31.6	25.7	—
Using cash	99.2	60.9	94.5	79.6	99.0	93.2	—
Using an account at a financial institution	6.2	44.0	2.9	39.7	0.2	8.7	—
Using a mobile phone	0.9	21.0	0.0	11.1	0.0	0.9	—
Paid utility bills							
In the past year	39.4	33.6	55.2	37.0	62.9	86.2	60.4
In the past year, rural	34.8	31.0	56.2	33.9	62.9	89.6	57.6
In the past year, income, poorest 40%	29.3	13.7	43.5	31.4	67.0	88.6	56.7
Using cash	99.7	76.3	98.1	87.1	98.8	98.5	78.6
Using an account at a financial institution	8.7	17.1	1.9	32.9	1.8	2.0	27.7
Using a mobile phone	0.5	55.2	0.6	7.9	0.1	0.9	3.4
Received government transfers							
In the past year	9.8	11.8	17.3	34.2	10.1	22.6	13.4
In the past year, rural	11.4	12.3	20.7	39.5	10.4	22.4	14.9
In the past year, income, poorest 40%	7.6	12.9	21.8	38.5	12.3	24.6	16.2
In cash	—	14.3	63.2	29.2	49.0	66.6	36.5
Into an account at a financial institution	—	51.7	23.0	82.0	52.5	39.6	60.7
Through a mobile phone	—	6.5	1.0	6.9	0.0	0.4	1.3
Received payments for agricultural products							
In the past year	20.9	53.7	22.1	10.7	20.3	36.8	—
In the past year, rural	22.3	56.8	31.5	14.2	22.6	41.7	—
In the past year, income, poorest 40%	28.0	55.3	25.3	11.2	24.3	42.6	—
In cash	84.5	94.1	95.4	76.0	97.4	98.9	—
Into an account at a financial institution	11.8	12.5	1.7	35.6	5.6	8.8	—
Through a mobile phone	2.3	30.4	0.2	11.6	0.0	0.7	—
Received wages							
In the past year	19.2	27.5	35.1	33.2	22.3	24.7	32.4
In the past year, rural	18.9	24.6	34.2	29.5	22.9	23.5	27.6
In the past year, income, poorest 40%	14.4	14.7	29.6	25.0	17.3	18.9	25.6
In cash	86.2	56.9	82.0	37.9	69.7	74.0	50.1
Into an account at a financial institution	20.1	48.7	18.0	79.1	32.1	33.6	54.3
Through a mobile phone	1.5	25.5	0.0	11.7	0.0	0.4	1.2

Source: 2014 Global Findex Survey data, http://datatopics.worldbank.org/financialinclusion/.

Note: — = not available. Not all questions were asked in all countries. The 2014 Global Findex Survey did not calculate global averages for these indicators.

Although promising outcomes are observed in each of the cases, the heavy presence of cash as the dominant choice of payments highlights the persistent challenges at the entry level of getting people to fully embrace e-money as the pathway to a fuller array of financial services including savings, credit, and insurance.

Highlights of Country-Specific Comparative Data

Kenya's Findex data show the dominance of mobile-phone–based accounts and increased financial access driven by M-Pesa (appendix A, table A.3). The share of adults with a transaction account is high at nearly 75 percent,[10] and the shares of women and the poorest 40 percent who hold accounts are around 71 percent and 63 percent, respectively. The share of the adult population with mobile accounts is 58 percent. These proportions are substantially higher than the averages for Sub-Saharan Africa (11.5 percent), for low-income countries (10 percent), and even for the world (2 percent). The shares of remittances received and sent by mobile phone (89 percent and 92 percent, respectively) are comparatively higher than Sub-Saharan African averages (28 percent and 31 percent, respectively). The receipt and payment numbers confirm that, in Kenya, people prefer mobile money as a transaction method, while the use of cash still remains high. Government payments favor financial institutions, indicating scope for further expansion of mobile services.

Findex data for South Africa clearly highlight the use of card payments and the link of government grants to debit cards, with debit card ownership of around 55 percent (appendix A, table A.5). Use of debit cards is also high: 41 percent of adults use them for payments, substantially higher than world and upper-middle-income country averages (23 percent and 20 percent, respectively). The shares of South Africans who send and receive remittances using money transfer operators are also high (57 percent and 61 percent, respectively). Government grants are channeled through financial institutions on bank-based debit cards. Wage payments go through financial institutions at a higher rate (27 percent) than in upper-middle-income countries and the world average (18 percent). Mobile money accounts are possessed by only 14 percent of South Africa's population, and cash still dominates payment transactions. Thus conditions are favorable for e-money to facilitate greater use of a wider range of financial services.

Sri Lanka's financial inclusion indicators are impressively high across all segments of the population (appendix A, table A.6). Even among the poorest 40 percent, 80 percent are included—far ahead of the South Asia, lower-middle-income country, and world averages (38 percent, 33 percent, and 54 percent, respectively). Although Sri Lanka has launched an innovative, interoperable mobile money solution, neither the inclusion numbers nor transaction volumes are driven by mobile money: only 0.1 percent of all adults have a mobile account. Financial institutions have provided accessibility to all segments of the population; hence mobile money is simply another option. Another factor is that mobile money was launched only recently (in 2012). In Sri Lanka, too, it is apparent that cash still plays a significant role and thus provides an opportunity for digital payments to expand.

Thailand also scores high on inclusion for all segments of the population: 78 percent of all adults, 75 percent of women, and 72 percent of the poorest 40 percent (appendix A, table A.7).[11] ATM transactions are especially high: more than 62 percent of adults identify ATMs as their main mode of withdrawal, compared with the upper-middle-income country average of 56 percent. Nevertheless, people use ATMs mainly for cash withdrawal, even though banks have enabled these customized ATMs to also handle utility, travel, and person-to-person payments at relatively low cost. Although mobile money is available in Thailand, it is not used to a significant degree: only 1.3 percent of all adults have a mobile account. Debit card ownership is fairly high, at 55 percent, but debit card *use* (8 percent) is lower than the East Asia and Pacific, upper-middle-income country, and world averages (15 percent, 20 percent, and 23 percent, respectively).

In India, cash dominates all types of transactions, and financial inclusion numbers are comparatively low (appendix A, table A.1). However, the impact of the recent digital financial inclusion drive in India (described in Part II of this volume, chapters 4 and 6) is evident when the 2011 and 2014 numbers are compared: adults with accounts increased from 35 percent in 2011 to 53 percent by 2014. This impressive growth bodes well for achievement of the goals for UFA2020. However, the transaction comparison shows cash dominance in all areas, representing a challenge for the government and financial institutions alike. The share of adults with mobile accounts is 2.4 percent, comparing favorably to regional and world averages (2.6 percent and 2 percent, respectively), although transactions using mobile money are negligible (except for receipts of payments for agriculture products). Encouraging people to move away from cash would be a priority for India.

The Philippines is included as a case study because mobile money initiatives originated in the Philippines in 2001, long before Kenya's M-Pesa mobile money platform started in 2007. Given the constraints on brick-and-mortar financial institutions due to the archipelagic nature of the country, one would have expected mobile money to flourish. However, only 4 percent of adults have mobile money accounts (appendix A, table A.4). While this rate is higher than the regional and lower-middle-income country averages (0.4 percent and 2.5 percent, respectively), it compares poorly with the 28 percent of adults who have bank accounts. The Philippines' total rate of financial inclusion—31 percent—is less than half the strong rate of financial inclusion in East Asia and the Pacific, where adult inclusion averages 69 percent. Remittance numbers are higher (34 percent receiving and 21 percent sending), but these are undertaken mainly through money transfer operators (mostly pawnshops). All transaction types reveal cash dominance. The discussion of the Philippines case study in Part II of this volume (chapters 4 and 6) will highlight the reasons why mobile money failed to gain traction.

The case of Indonesia, as noted earlier, represents achievement of interoperability to help overcome geographic challenges to inclusion. On most financial indicators, Indonesia remains below average in the region despite substantial progress between 2011 and 2014 (appendix A, table A.2). The share of adults with an account at a financial institution rose from 20 percent in 2011 to 36 percent in 2014, still below the averages of 42 percent for lower-middle-income countries and 69 percent for East Asia and the Pacific.

Saving at a financial institution likewise rose from 15 percent to 27 percent, and the share of respondents with a debit card rose even more sharply from 11 percent to 26 percent—comparable to lower-middle-income countries, although below the regional average. Digital transactions and use of mobile phones and money for remittances also are below average, although Indonesians make relatively high use of ATMs, which are the main mode of withdrawal for 71 percent of those with an account.

Maldives is not covered by Findex data. However, previous World Bank engagements show that the country's archipelagic nature makes it difficult to provide financial access to the 340,000-plus people living among more than 200 islands. Over half of the population is in the outer islands, and those living on atolls have great difficulties accessing banking and payment services. Although banking services are woefully inadequate, a World Bank–funded mobile-phone banking project closed without being able to successfully deploy the much-needed mobile solution. The World Bank and the local Maldives authorities are trying to address this issue through a different payment solution. The case study (in chapter 4) highlights important lessons learned from the past difficulties and problems encountered and the nonbank-based approach being undertaken.

Organization of This Volume

Following this "Overview" chapter, Part I of the volume provides the framework for studying the journey toward a cash-lite society and financial inclusion, as follows:

- *Chapter 1, "The Challenge of Financial Inclusion,"* introduces the topic of financial inclusion, discusses its importance, and presents definitions.
- *Chapter 2, "Digitizing Financial Inclusion through Innovations,"* explains how digitizing money, payments, and other financial transactions can facilitate inclusion of the unbanked and financially underserved population. It also addresses the risks of digital finance, including
  - *Political economy risks* that stem from vested interests that resist the introduction of disruptive technologies;
  - *Security risks and risks to the payments system,* which may not function effectively if the service providers do not adequately understand the digital customer base or the infrastructure requirements;
  - *Principal-agent risks* that must be addressed to minimize loss of revenue, fines and other reprimands by the regulators, fraudulent activities and corruption, and loss of reputation; and
  - *Risks to customers* from inadequate information and understanding of e-money, lack of consumer protection and adequate redress mechanisms, identity theft, and liquidity-related issues, among others—all of which may account in part for the trust gap that impedes increased and broader use of

digital finance and the achievement of higher levels of financial inclusion beyond entry-level transactional and payment services.

- *Chapter 3, "Stakeholders in Digital Financial Inclusion,"* sets forth the key stakeholders in the process of applying digital innovations to achieve greater financial inclusion.

Part II presents the empirical evidence on critical enablers that are game changers in successful e-money deployments, as follows:

- *Chapter 4, "Policy Leadership and Enabling Regulatory Environments,"* focuses on the macro environment, including leadership in policy and regulatory reforms that are both successful and less successful in achieving results in terms of financial inclusion.
- *Chapter 5, "Innovative Uses of Infrastructure and Digital Ecosystems,"* examines measures at the meso level (the institutional framework) to establish enabling infrastructure, institutions, and social grant payment systems that are essential alternative payment system platforms designed to drive demand and use by the poor and unbanked. In chapter 5, particular attention is paid to the micro-level private service providers and how they have responded and rapidly built up agent networks. Kenya's unusually rich set of mobile money applications and private development of the digital ecosystem have allowed nonbank provision of payment services to include the poor in an array of credit, savings, insurance, and other financial products.
- *Chapter 6, "Unique Identification,"* looks at different experiences with national identification (ID) systems, because a unique national ID is found to be a critical condition to link e-money platforms and other digital solutions to bank accounts in order to transform financial inclusion of the poor, rural, and unbanked.

Part III draws on the lessons of experience to distill guiding principles for success and measures that would help avoid deficiencies that have constrained the pace of scaling-up in several South Asian countries, as follows:

- *Chapter 7, "Digital Landscape in South Asia,"* summarizes the e-money landscape across countries in the South Asia region. Although such an overview cannot replace in-depth diagnostic assessment of how best to digitize the delivery of financial services in each country, it provides a basis for the discussion in chapter 8.
- *Chapter 8, "Opportunities, Challenges, and Future Options in South Asia,"* suggests key options that could be applied in South Asian countries.
- *Chapter 9, "Conclusions,"* reviews the conclusions regarding the critical enablers and game-changing measures for digital applications to help transform financial inclusion in developing countries more generally.

Notes

1. Throughout this volume, "developing countries" refers to low- and middle-income economies as defined in the World Bank's Income Classifications, based on estimates of gross national income (GNI) per capita, calculated using the World Bank Atlas method. As of July 1, 2016, low-income economies are those with 2015 GNI per capita of US$1,025 or less. Middle-income economies are those with 2015 GNI per capita of US$1,026 to US$12,475 ("upper-middle-income" economies having 2015 GNI per capita of US$4,036 to US$12,475). "Developed countries" refers to high-income economies, which are those with 2015 GNI per capita of US$12,476 or more. Classification by income does not necessarily reflect development status. For more information, see https://datahelpdesk.worldbank.org/knowledgebase/articles/906519-world-bank-country-and-lending-groups.
2. Enterprise Surveys refer to firm-level surveys of a representative sample of an economy's private sector. The surveys cover a broad range of business environment topics including access to finance, corruption, infrastructure, crime, competition, and performance measures. Since 2002, the World Bank has collected these data from face-to-face interviews with top managers and business owners in over 155,000 companies in 148 economies. For more information about the Enterprise Surveys, see http://www.enterprisesurveys.org/.
3. For more about the UFA2020 initiative, see the overview brief on the World Bank website: http://www.worldbank.org/en/topic/financialinclusion/brief/achieving-universal-financial-access-by-2020.
4. Data also from Global Findex 2014 Survey, http://datatopics.worldbank.org/financialinclusion/. The name of "M-Pesa," Kenya's mobile-phone–based branchless banking service, is derived from M for mobile and "pesa" (Swahili for money).
5. South Africa has issued 10 million cards covering 16 million beneficiaries.
6. Data from Global Findex 2014 Survey, http://datatopics.worldbank.org/financialinclusion/.
7. The total inclusion figure differs from that in the Global Findex 2014 data (appendix A, table A.7) because the BOT (2014) methodology differs from that of Findex. Among other differences, 7.8 percent of households that *voluntarily* do not use financial services are not considered "excluded" in the BOT survey, so they are implicitly counted within the BOT (2014) "inclusion" figure.
8. Findex data come from the Global Findex Database, http://datatopics.worldbank.org/financialinclusion/. The data are compiled using the Gallup World Poll Survey and measure how adults in 143 economies around the world manage their day-to-day finances and plan for the future.
9. Also treated in this study (chapter 5) along with South Africa as exemplary social grant transfer payment programs are Mexico's Oportunidades and Brazil's Bolsa Família programs, which were considered for in-depth case study fieldwork. The choice was made to pursue fieldwork and in-depth treatment of South Africa's biometric chipped debit card program because it is considered a more flexible e-money product as an effective entry point into the financial system by the poor. That is, poor grant recipients can use their chipped card at any bank ATM or point of sale (POS). By comparison, the prepaid cards used in Mexico and Brazil are "closed loop" and are primarily used for identifying the grant beneficiary for cash-in/cash-out transactions at limited locations.

10. Figure may differ from other data sources depending on definitions and methodology. FSD Kenya and CBK (2013) provide a 70 percent figure for overall financial inclusion in Kenya.
11. Figures depend on definitions and methodology. For Thailand, for example, BOT (2014) reports somewhat higher figures for financial inclusion than the Findex Survey 2014.

Bibliography

BFA (Bankable Frontier Associates). 2015. "The Journey toward 'Cash-Lite': Addressing Poverty, Saving Money and Increasing Transparency by Accelerating the Shift to Electronic Payments." BFA study for the Better than Cash Alliance, Somerville, MA.

BOT (Bank of Thailand). 2014. "Financial Access Survey of Thai Households 2013." Survey report, Financial Institutions Strategy Department, BOT, Bangkok.

Cull, R., T. Ehrbeck, and N. Holle. 2014. "Financial Inclusion and Development: Recent Impact Evidence." Focus Note 92, Consultative Group to Assist the Poor (CGAP), Washington, DC.

Demirgüç-Kunt, A., T. Beck, and P. Honohan. 2008. "Finance for All? Policies and Pitfalls in Expanding Access." A World Bank Policy Research Report, World Bank, Washington, DC.

FinMark Trust. 2014. "FinScope South Africa 2014." Annual FinScope consumer survey summary, FinMark Trust, Johannesburg.

FSD (Financial Sector Deepening) Kenya and CBK (Central Bank of Kenya). 2013. "FinAccess National Survey 2013: Profiling Developments in Financial Access and Usage in Kenya." Survey results report, FSD Kenya and CBK, Nairobi.

Hawkins, D. 1980. *Identification of Outliers*. London, New York: Chapman and Hall.

Sahay, R., M. Čihák, P. N'Diaye, A. Barajas, R. Bi, D. Ayala, Y. Gao, et al. 2015. "Rethinking Financial Deepening: Stability and Growth in Emerging Markets." IMF Staff Discussion Note SDN/15/08, International Monetary Fund, Washington, DC.

Tambunlertchai, K. 2015. "Financial Inclusion, Financial Regulation, and Financial Education in Thailand." ADBI Working Paper 537, Asian Development Bank Institute, Tokyo.

World Bank. 2014. *Global Financial Development Report 2014: Financial Inclusion.* Washington, DC: World Bank.

———. 2017. *Doing Business 2017: Equal Opportunity for All.* Washington, DC: World Bank.

PART I

Journey toward a Cash-Lite Society and Financial Inclusion

Part I introduces the topic of financial inclusion (chapter 1), discusses its importance, presents definitions, and explains how digitizing money, payments, and other financial transactions can facilitate inclusion of the unbanked and financially underserved population (chapter 2). It also sets forth the key stakeholders in the process of applying digital innovations to achieve greater financial inclusion (chapter 3).

CHAPTER 1

The Challenge of Financial Inclusion

What Is Financial Inclusion?

This chapter introduces the concept of financial inclusion and why it matters for development. The gaps in financial inclusion in South Asian countries are analyzed along several dimensions and are related to poverty status. Particular attention is paid to the role of remittance transfers.

Financial inclusion has gained greater prominence in recent years as a key priority in the reform and development agenda to achieve sustainable and inclusive economic growth. In 2009, the Group of Twenty (G20) included financial inclusion on its agenda (G20 2009); since that time, donor communities, standard-setting bodies, national-level policy-making and regulatory bodies, and academia have embarked on various initiatives to further financial inclusion.

The benefits of financial inclusion are widely accepted. This volume does not aim to undertake a full-fledged literature review to explain the theoretical underpinnings of financial inclusion or evaluate the development benefits that would accrue from having inclusive, low-cost, accessible, and reliable financial systems. However, the relevance of financial inclusion to the developing world, especially to the South Asian countries that motivated this study, is described below with the financial inclusion gap as the axis for the analysis.

The Consultative Group to Assist the Poor's (CGAP 2011) working definition of financial inclusion (box 1.1) explains the need to address access, use, and quality of service dimensions in order for a financial system to be truly inclusive. Two other important points are also highlighted: that certain groups are excluded by the financial system, and that exclusion happens in payments as well as in traditional intermediary markets such as savings, credit, and insurance.

Box 1.1 "Financial Inclusion": A Working Definition

"Financial inclusion" refers to a state in which all working-age adults, including those currently excluded by the financial system, have effective access to the following financial services provided by formal institutions: credit, savings (defined broadly to include current accounts), payments, and insurance.

"Effective access" involves convenient and responsible service delivery, at a cost affordable to the customer and sustainable for the provider, with the result that financially excluded customers use formal financial services rather than existing informal options.

"Financially excluded" refers to those who do not have access to or are underserved by formal financial services.

"Responsible delivery" involves both responsible market conduct by providers and effective financial consumer protection oversight.[a] The specific characteristics of excluded consumers have significant implications for effective consumer protection regulation and supervision, and therefore also for standards and guidance aimed at enabling financial inclusion. Relevant characteristics are likely to include limited experience with, and sometimes distrust in, formal financial service providers; lower levels of education and financial literacy and capability; few formal providers to choose from, if any; and remote locations.

"Formal institution" refers to a financial service provider that has a recognized legal status and includes entities (and, in some countries, even some individuals) with widely varying regulatory attributes, subject to differing levels and types of external oversight. However, the fact that a customer's financial service provider has a recognized legal status does not mean she or he should be considered "financially included" under the definition used: for this, all the conditions of "effective access" must be met. Moreover, formal products and providers do not in all cases offer customers a better value proposition than informal products and providers. The reality for many financially excluded households is that informal options may be the best they have available for the foreseeable future for at least some of their financial service needs.

Source: CGAP 2011.
a. "Responsible market conduct" by providers includes reasonable steps to ensure transparency and fair treatment, and to mitigate consumer risks.

According to Kempson and Whyley (1999a, 1999b), there are five major forms of exclusion:

- *Access exclusion*, where segments of the population remain excluded from the financial system because of either remoteness or the process of risk management of the financial system
- *Condition exclusion*, when exclusion occurs because of conditions that are inappropriate for some people
- *Price exclusion*, when financial products are unaffordable

- *Marketing exclusion*, when targeted marketing and sales of financial products lead to exclusion
- *Self-exclusion*, which takes place when certain groups of people exclude themselves from the formal financial system owing to fear of refusal or other psychological barriers

Financial inclusion, therefore, is not merely the outcome of a process, but a process in itself. Because of its multidimensional nature, it remains hard to interrelate such dimensions over time. As a self-reinforcing cycle that results from the accumulation of a number of disadvantages, it is difficult to attribute causality to one specific factor or another.

Recognizing the multidimensionality of financial inclusion has led to changes in the way it is measured. Similarly, an understanding has emerged that financial inclusion is not static, but rather dynamic, and that different individuals or groups find themselves in different stages of the financial inclusion process, at times temporarily, recurrently, or continuously. Moreover, financial exclusion can occur in one or more of the essential markets, that is, transaction banking, savings, credit, and insurance. One can therefore argue that no single intervention can address all of these complexities and achieve the desired state of inclusiveness. Hence, even well-developed financial markets have their own financially excluded segments of the population. Developing countries should assess the rapidly changing financial and social landscapes and focus on whether vulnerable groups face increased risks of financial exclusion.

Although the measurement of financial inclusion using simplistic indicators is adequate for assessing the size of the exclusion problem, it may not be sufficient to properly inform policy makers in these developing countries; as such, the best means of measuring inclusion is still under debate. A literature review reveals that different approaches have been proposed, including the use of a variety of financial inclusion dimensions, measurable proxies, and a few econometric estimations. Because consensus is lacking about a better indicator that can be easily measured, all available datasets (Findex 2011 and 2014 Surveys;[1] Honohan 2008; Sophastienphong and Kulathunga 2010) use banking inclusion (both demand- and supply-side) as analogous to financial inclusion.

In the second Global Findex Survey (2014), a more rounded understanding of financial inclusion is developed to identify opportunities to remove barriers that may prevent people from using financial services. Historically, financial inclusion was associated with the branch banking model, and therefore almost all measurements of financial inclusion were based on banking density, proximity, availability of banking facilities, and affordability of banking products. The 2014 Findex Survey revised this narrow view. By adding digital platforms, such as mobile money accounts (using mobile phones), to its financial inclusion measures, a wider array of indicators (in particular, payment service indicators) are recognized, allowing for more realistic interpretations and comparisons. Thus, for the first time, the contribution of digitized payments in furthering financial inclusion has been measured globally.

It should be noted that this measurement of inclusion is still an evolving process. Policy makers must often take a broader view of economic and financial development by observing worldwide trends, innovations, and the full range of policy initiatives that they might use to design, implement, and monitor progress for the promotion of financial inclusion.

Why Does Financial Inclusion Matter?

According to the 2014 Global Findex Survey,[2] around 38 percent of adults (ages 15 and older) living in low- and middle-income economies cannot access basic services such as bank accounts, bill payment facilities, mobile money accounts, loans, insurance, and other simple financial services because of excessive costs, travel distance, documentation needs, and even lack of awareness. The majority of this population is at the base of the pyramid and hence is highly vulnerable to being unable to cope with unexpected events, such as sickness or death of the breadwinner, loss of employment, crop failure, and natural calamities.

Income and expenditure shocks foster disruptive financial practices in an effort to smooth consumption, such as borrowing from loan sharks at prohibitive rates and selling off assets. As a result, the poor are often unable to climb out of the poverty trap. Access to the financial system provides fairer opportunities for those living in poverty to improve their income and their standard of living. As stated by the Commission on Growth and Development (CGD 2008), if the financial system fails to reach large portions of the population, household savings will be stunted. People need a secure, accessible vehicle in which to store their wealth. If the banks do not provide such vehicles, people will save less or will simply store their money in less liquid forms that do not serve well the wider economy.

The Link between Financial System Growth and Social Inclusion

In recent literature, economists agree on the positive impacts of financial sector development on economic growth. These are well documented in studies by Creane et al. (2004) and Merton and Bodie (1995) on the ways in which development of the financial system enhances the efficiency of intermediation and results in better resource allocation; in Levine's (2005) seminal paper on the merits of financial sector development; and in more recent work by Claessens and Feijen (2006) and Ahmad and Malik (2009).

Well-functioning financial systems not only play a critical role in sustaining paths of high economic growth by mobilizing savings from the public and allocating the funds to productive investments; they also help make growth more inclusive by providing access to finance for all, which is associated with more rapid growth and job creation, better income distribution, and poverty reduction (Kulathunga 2012).

Financial intermediation is a critical factor for growth and social inclusion. One of its core functions is to mobilize financial resources from surplus agents and channel them to those with deficits. It thus allows investor entrepreneurs to expand economic activity and employment opportunities. It also enables household consumers, as well as micro- and small-business entrepreneurs, to expand their own welfare and earnings opportunities and seek to smooth their lifetime outlays. In all cases, financial intermediation drives economic growth and contributes to social inclusion, provided it is conducted in a sound and efficient way (Grais and Kulathunga 2007). Hence, financial intermediation is a key link between a country's overall economic growth and its level of social inclusion.

In underdeveloped financial systems, access is limited. People must resort to high-cost informal sources, either because formal financial intermediaries are geographically unavailable or because their products are inaccessible because of cost and distance. This constrains participation in economic activities, affecting growth patterns spatially and across social and economic levels. In contrast, a key feature of financial deepening is that it accelerates economic growth by expanding access to those who lack adequate finance.

The Link between Financial Inclusion and Poverty Reduction

Access to financial services provides individuals with the opportunity to manage risk, broaden their menu of economic choices, and smooth their consumption. Increasing financial access provides capital for enterprise expansion, protects against both covariate and idiosyncratic shocks, helps move money between family members across the world, and generally improves the well-being and economic sustainability of the poor.

Theoretically, this promotes economic development, thereby contributing to poverty reduction. For example, Beck, Demirgüç-Kunt, and Levine (2007) used cross-country data to show that financial development disproportionately raises the incomes of the poorest quintile, both by directly reducing income inequality and, more powerfully, through impacts on aggregate economic growth.

In their paper introducing the Global Findex Database 2014, Demirgüç-Kunt et al. (2015) describe how policy makers and regulators are increasingly making financial inclusion a priority in financial sector development. Accordingly, 67 percent of 147 jurisdictions have a mandate to promote financial inclusion, and international organizations—including the G20 and the World Bank—are also beginning to formulate strategies that promote financial inclusion.

The Benefits of Financial Inclusion: The Evidence

It is also important to assess the evidence that inclusive and efficient financial markets have the potential to improve the lives of citizens, reduce transaction costs, spur economic activity, and improve delivery of other social benefits and innovative private sector solutions. A 2014 CGAP Focus Note examines the evidence and provides a comprehensive evaluation attesting to the positive impacts (Cull, Ehrbeck, and Holle 2014):

- At the microeconomic level, small businesses benefit from access to credit, while the impact on the broader welfare of the borrower's household might be more limited.

- Savings help households manage cash flow spikes, smooth consumption, and build working capital.
- Access to formal savings options can boost household welfare.
- Insurance can help poor households mitigate risk and manage shocks.
- New types of payment services can reduce transaction costs and seem to improve a household's ability to manage shocks through shared risks.
- Financial access improves local economic activity: at the macroeconomic level, empirical evidence shows that financial inclusion is positively correlated with growth and employment.
- Financial inclusion can improve the effectiveness and efficient execution of government payment of social safety net transfers (government-to-person, or G2P, payments), which play an important role in ensuring the welfare of many poor people.
- Financial innovation can significantly lower transaction costs and increase reach, which is enabling new private sector business models that help address other development priorities.

The Global Financial Inclusion Gap

Financial inclusion is a global challenge. Inclusive financial systems—allowing broad access to appropriate financial services—are especially likely to benefit poor people and other disadvantaged groups (Demirgüç-Kunt and Klapper 2013). The Global Findex 2014 data show that 54 percent of the adults from the world's poorest 40 percent of households do not have accounts, not only because of poverty, but also because of prohibitive costs, travel distance, and the paperwork involved.[3]

Globally, the financial inclusion gap, according to Findex 2014, represents around 38 percent of the total adult population. According to Findex 2011 data—in which financial inclusion had been measured based on the number of adults having accounts only in formal financial institutions—the comparable global financial access gap was around 50 percent. Thus, adding and measuring mobile money transaction accounts in the Findex 2014 Survey helped to narrow the inclusion gap between 2011 and 2014 and also portrays a more complete picture.

Before the publication of the first Findex Survey data in 2011, Honohan (2008) estimated the fraction of the adult population using formal and semiformal financial services (that is, from unregulated microfinance institutions) for more than 160 countries by combining data from banks and microfinance institutions with household surveys. Matching this information with 2005 population data as a basis, the financially excluded population was estimated at around 54 percent of the total adult population in *Banking the World: Empirical Foundations of Financial Inclusion* (Cull, Demirgüç-Kunt, and Morduch 2013). Honohan's estimate (table 1.1) is reasonably consistent with Findex data. It reveals that financial service use in low- and middle-income countries—which averages around 37 percent—is woefully low compared

Table 1.1 Estimated Financial Inclusion Gap, Globally and by Region, 2008

Region	*Adult population (millions)*	*Using financial services (millions)*	*Not using financial services (millions)*	*Share using financial services (%)*
Middle East and North Africa	191.70	63.28	128.42	33
Central Asia and Eastern Europe	381.87	195.00	186.87	51
East Asia	1,471.20	607.90	863.29	41
High-income OECD	588.94	539.65	49.29	92
Latin America and the Caribbean	387.13	136.82	250.31	35
South Asia	1,039.50	435.76	603.74	42
Sub-Saharan Africa	370.02	74.43	295.60	20
Total	4,430.35	2,052.83	2,377.52	46

Sources: Cull, Demirgüç-Kunt, and Morduch 2013, based on Honohan 2008.
Note: OECD = Organisation for Economic Co-operation and Development. "Adult population" is defined as ages 15 years and older. "Financial services" includes both formal and semiformal financial services (that is, from unregulated microfinance institutions but not from mobile money accounts).

with high-income Organisation for Economic Co-operation and Development (OECD) countries, where 92 percent of adults use financial services. The rate of exclusion in South Asia is roughly 58 percent, while that of Sub-Saharan Africa is highest, with 80 percent of the adult population in the excluded category.

South Asia's Financial Inclusion Gap

The South Asia region is home to many of the developing world's poor. According to the World Bank's most recent poverty estimates for 2012, about 309.2 million people in the region survive on less than US$1.90 a day (the reestimated extreme poverty line),[4] and they make up more than 34.3 percent of the developing world's poor (World Bank 2016).

Thanks to robust economic growth averaging 6 percent per year over the past 20 years, South Asia has experienced declining poverty and impressive improvements in human development. The percentage of people living on less than US$1.25 a day (the extreme poverty line prior to 2015) fell in South Asia from 61 percent to 24.5 percent between 1981 and 2011 (World Bank 2015a). At US$1.90 a day, this was projected to drop further, to 13.5 percent, by 2015 (World Bank 2016).

However, factors such as weak global demand and lower foreign investment pursuant to the global financial crisis, political uncertainties, power outages, and weak monsoons resulted in the slowing of economic growth in the region to around 4.2 percent in 2012, from an estimated 7.2 percent in 2011. If these volatilities are addressed effectively, it is expected that South Asia can grow by an average of around 6.5 percent over the next three years. Nevertheless, the slowdown in global growth patterns, conflicts, and volatile markets are continuing challenges.

Figure 1.1 Share of Adults with a Financial Services Account, by Region, 2014

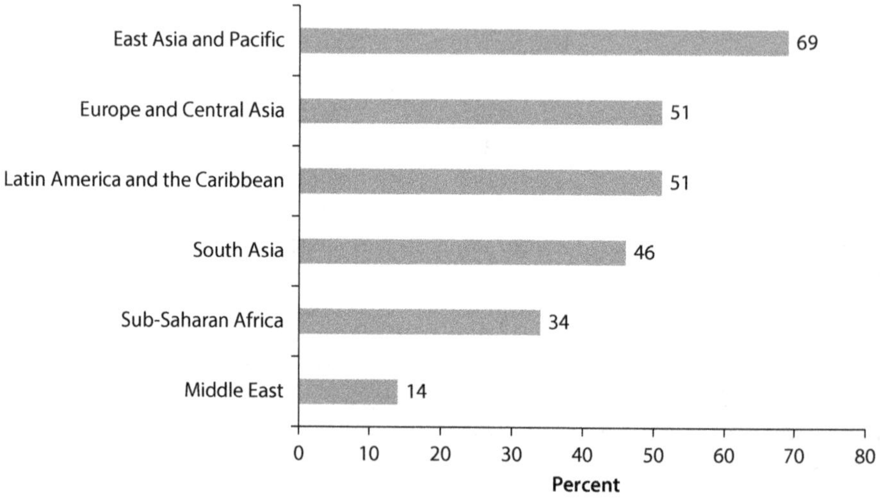

Source: Global Findex Survey 2014 data, http://datatopics.worldbank.org/financialinclusion/.
Note: Adults are those aged 15 years or older. "Financial services" includes both services from formal financial institutions and mobile money accounts.

In terms of financial access, South Asia ranks above the Middle East and Sub-Saharan Africa regions, with around 46 percent of the adult population having accounts (figure 1.1).[5] Thus the financial inclusion gap—measured in terms of the number of adults without an account in a formal financial institution or a mobile money account as the proxy—is around 54 percent.[6]

In recent times, retail payment systems development in Asia as a whole has been impressive, although growth rates in South Asia have been low relative to those in East Asia (China, Japan, and the Democratic People's Republic of Korea) and in Southeast Asia (Indonesia, Malaysia, Singapore, and Thailand). It is also noted that technology-driven retail payment services have clearly demonstrated their contribution to reducing poverty (especially in East Asian and Southeast Asian countries) by ensuring the availability and affordability of financial services—in other words, financial inclusion.

However, until recently, no meaningful correlation has been established between payment services and poverty reduction, although some preliminary attempts to establish such a relationship are now apparent (as in Cull, Demirgüç-Kunt, and Morduch 2013). As such, the potential of retail payment services as a poverty reduction measure has still not been explored fully, primarily because of the narrow range of indicators used to measure poverty reduction as well as the labeling of payment services as an information and communication technology (ICT) or information technology (IT) enabler rather than as a business proposition or a potential measurement of the social well-being of the people. Ideally, this perception will change with the reclassification of the definition of account ownership to include mobile money accounts, with the 2014 Findex data as a starting point.

Country-Specific Inclusion Gaps

Analysis of country data shows that almost all South Asian countries fall below the regional average (46 percent) for financial inclusion except for Sri Lanka, which has 83 percent inclusion.[7] At roughly 10 percent and 13 percent, respectively, Afghanistan and Pakistan register the lowest rates in the region (figure 1.2). In South Asia overall, 54 percent of adults, or around 585 million people, remain outside of the formal financial system.

In the developing world as a whole, it is interesting to note that lack of money is the most commonly reported reason for being unbanked. Other reasons for not having access to formal financial services include distance to banks and the costs associated with maintaining an account at a formal institution such as a bank, credit union, savings and credit cooperative, post office, or microfinance institution (Demirgüç-Kunt and Klapper 2013).

Further analysis of South Asian country data shows a different dimension of the inclusion gap: gender-based inequality. Findex data highlight this inequality within and across gender: in all South Asian countries except Sri Lanka, fewer women than men have financial services accounts (figure 1.3). There is striking disparity in Afghanistan and Pakistan, where less than 4 percent of women have formal financial access compared with rates nearly four times greater for men, at 15.8 percent and 14.2 percent, respectively. Even in India, the rate for men (62.5 percent) is 47 percent higher than the rate for women (42.6 percent).

Figure 1.2 Share of South Asian Adults with a Financial Services Account, by Country, 2014

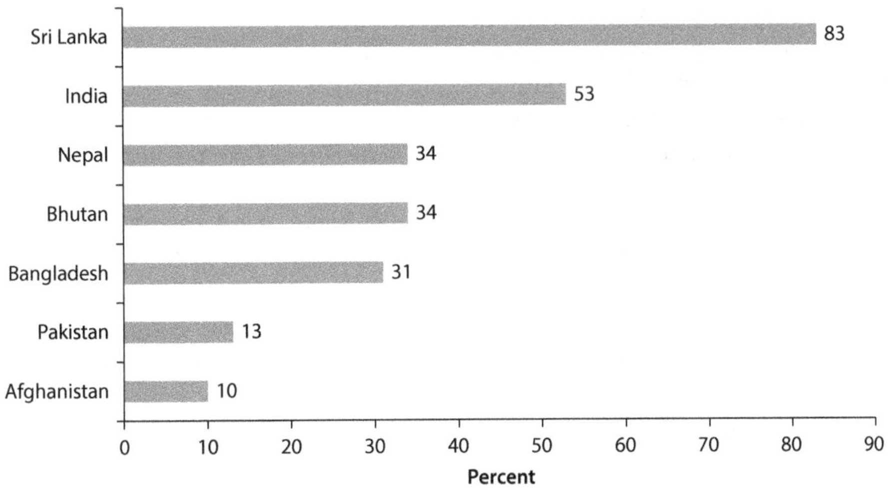

Source: Global Findex Survey 2014 data, http://datatopics.worldbank.org/financialinclusion/.
Note: Adults are those aged 15 years or older. "Financial services account" includes either an account with a formal financial institution or a mobile money account.

Figure 1.3 Share of South Asian Adults with a Financial Services Account, by Gender and Country, 2014

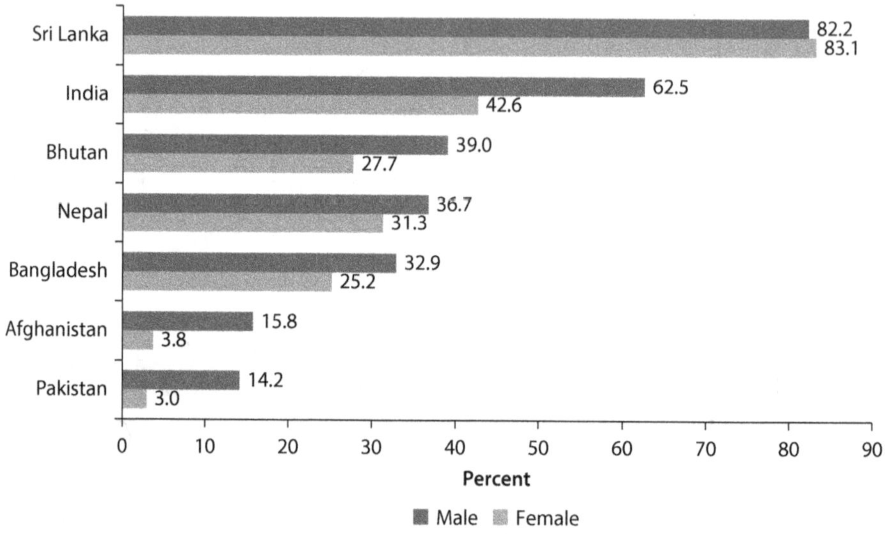

Source: Global Findex Survey 2014 data, http://datatopics.worldbank.org/financialinclusion/.
Note: Adults are those aged 15 years or older. "Financial services account" includes either an account with a formal financial institution or a mobile money account.

The Supply-Side Dimension: Finance Accessibility

Because Findex depicts only demand-side data and does not include Maldives, the study also reviews the factors in the supply-side dimension for all South Asian countries, using the Getting Finance in South Asia 2010 database (Sophastienphong and Kulathunga 2010). Figure 1.4 shows standardized data for South Asia for the period 2004–08. Maldives and Sri Lanka both register high scores on these indicators of accessibility of finance.

It should be mentioned that, although the penetration numbers for banking access are high for Maldives (in terms of geographic and demographic penetration of automated teller machines [ATMs] or branches), the ratios alone do not reflect the true picture, because geographic dispersion of the population among the atolls is not taken into consideration in the calculations. Most bank branches or ATMs are concentrated on the main island of Malé, while the Maldives population of 338,400 is scattered throughout 200 islands in 26 atolls. Most people who live on atolls are a two- to four-hour round-trip ferry ride away from the nearest bank branch or ATM; thus, to date, although the financial access figure is estimated at around 50 percent of the population, cost and proximity issues create untold hardships for Maldivians to access financial services.

India scores well on all indicators, while Pakistan's results suggest a need to focus on the use of financial services, notably the provision of loans and the mobilization of deposits (Sophastienphong and Kulathunga 2010). Bangladesh scores low on demographic penetration for both bank branches and ATMs, while

Figure 1.4 Access to Finance in South Asia: Supply-Side Data, 2010

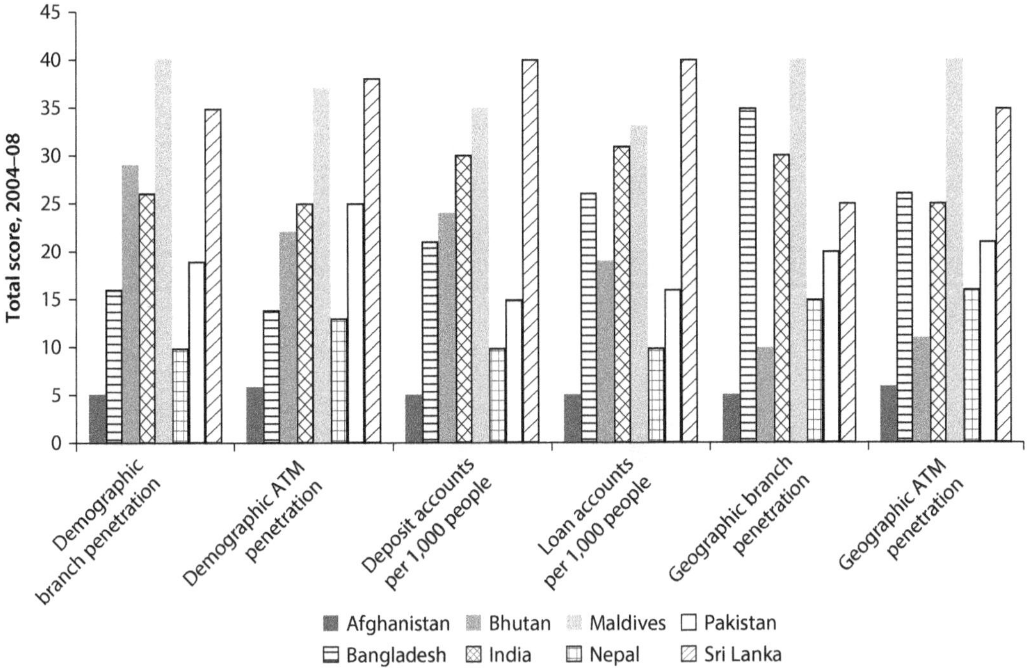

Source: Sophastienphong and Kulathunga 2010.
Note: ATM = automated teller machine. Scores are based on each country's position on a scale from 5 (lowest-performing country) to 40 (highest-performing country) for each indicator.

Bhutan scores low on geographic penetration. Nepal's scores suggest a need to work on all aspects of access to finance, though increasing the number of players in the market has not produced the desired effect. Afghanistan scores lowest on all indicators.

Poverty, Financial Exclusion, and Financial Vulnerability in South Asia

The overall findings on poverty, the financial inclusion gap, and financial vulnerability in South Asia are summarized in figure 1.5. According to Findex 2014 data, the majority of the adult population in South Asia (54 percent, or around 585 million people)[8] is financially excluded. According to World Bank (2015a) data, the population's extreme-poverty headcount ratio (below US$1.25 per person per day) in 2014 was around 24 percent (399 million people). Taken together, these data indicate that, in addition to the extreme poor, some 186 million people (around 11 percent of the South Asian adult population) are financially vulnerable though not in the extremely poor category.

Figure 1.5 Poverty, Financial Exclusion, and Financial Vulnerability Indicators in South Asia, 2014

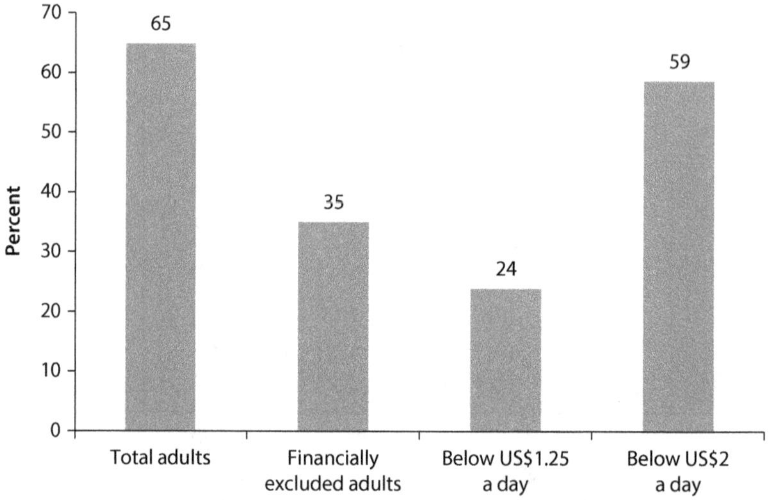

Source: Calculations based on Global Findex Survey 2014 data (http://datatopics.worldbank.org/financialinclusion/) and World Bank 2015a.
Note: "Financially excluded adults" refers to people ages 15 and older who lack any kind of account for financial transactions, either through a formal financial institution or a mobile money service. "US$1.25 a day" represents the 2014 international per capita "extreme poverty" line, using 2005 purchasing power parity (PPP) exchange rates. "Below US$2 a day" represents the median poverty line of all developing countries, also at 2005 PPP, as calculated by Chen and Ravallion (2010).

The fact that South Asia accounts for a third of the world's poor (World Bank 2016) indicates the challenge that these financial inclusion indicators pose for addressing poverty in the region. However, one needs to be prudent in interpreting these statistics or generalizing the results, as they may not have fully captured the more recent technology-based financial services that are being successfully delivered in some countries in South Asia, especially in Bangladesh, India, and Sri Lanka.

Age dependency on the working-age population is another important dimension that cannot be ignored, especially when considering a country's poor.[9] At the base of the pyramid, living with less money means having very little control over economic outcomes and virtually no ability to save for unforeseen events. The ramifications of financial exclusion for working-age adults inevitably pass on to their dependents. In South Asia, the age dependency ratios of the young (ages 0–14 years) and old (ages 65 years and older) are 46.4 percent and 7.9 percent, respectively.[10] Therefore, in South Asia, there are nearly 55 dependents per 100 working-age adults, on average. Thus, the total financially excluded population in South Asia can be considered to exceed 1.5 times the number of working-age adults identified as financially excluded.

The extent of concentration of the world's extremely poor in South Asia argues for policy actions and approaches to address the resulting financial inclusion gap. For example, World Bank (2016) data establish that in 2012, Bangladesh (at 43.6 percent, or 66 million people) and India (at 21.2 percent, or 268 million people) were among the top 10 countries with the largest numbers of extreme poor in the world.

At the same time, it is encouraging to see that India led the five top contributors to poverty reduction from 2008 to 2011 by lifting a staggering 140 million people out of extreme poverty (World Bank 2015a). These numbers tell us that, if appropriate policy responses and corrective actions are taken to close financial inclusion gaps in South Asia, these significant poverty figures can be further reduced by giving poor people more opportunities to participate in the economic development process in these countries. The World Bank's expectation is that the extreme poor population in South Asia will be further reduced to 249.6 million (35.8 percent of the world's extreme poor population) by 2020, and to a remarkable 42.5 million (10.3 percent of the world's extreme poor) by 2030 (World Bank 2015a). Nevertheless, policy makers and researchers have not been able to capture the full contribution to poverty reduction of technology-based payment services because of the limited availability of reliable information.

Remittance Transfers and Financial Inclusion

Introducing the Global Findex 2014 Database, Demirgüç-Kunt et al. (2015) highlight the importance of domestic remittances in developing economies. According to Findex Surveys, 15 percent of adults in developing economies reported having sent money to a relative or friend living in a different part of the country, and 19 percent report having received such a payment in 2014. While 48 percent of adults questioned in Sub-Saharan Africa reported having either sent or received remittances, in South Asia this figure reached only around 17 percent.

Although cash remains by far the most common method of sending or receiving domestic remittances, sending money digitally is gaining popularity, with mobile money being the preferred method. Hence remittances are examined here as a potential driver of increased use of mobile money, which in turn is an entry point for access to additional financial services.

Significance of Remittances to Developing Countries

For many developing economies, remittances are an important source of real convertible currency. On a macroeconomic level, remittances are more stable than other sources of foreign exchange because their variation is often countercyclical, helping to sustain consumption and investment during downturns. The value of remittance flows to developing economies is now more than three times that of official development assistance (figure 1.6). For countries such as Bangladesh, the increasing flow of remittances has reduced external aid dependency,

Figure 1.6 Remittances and Other Resource Flows to Developing Countries, 1990–2015

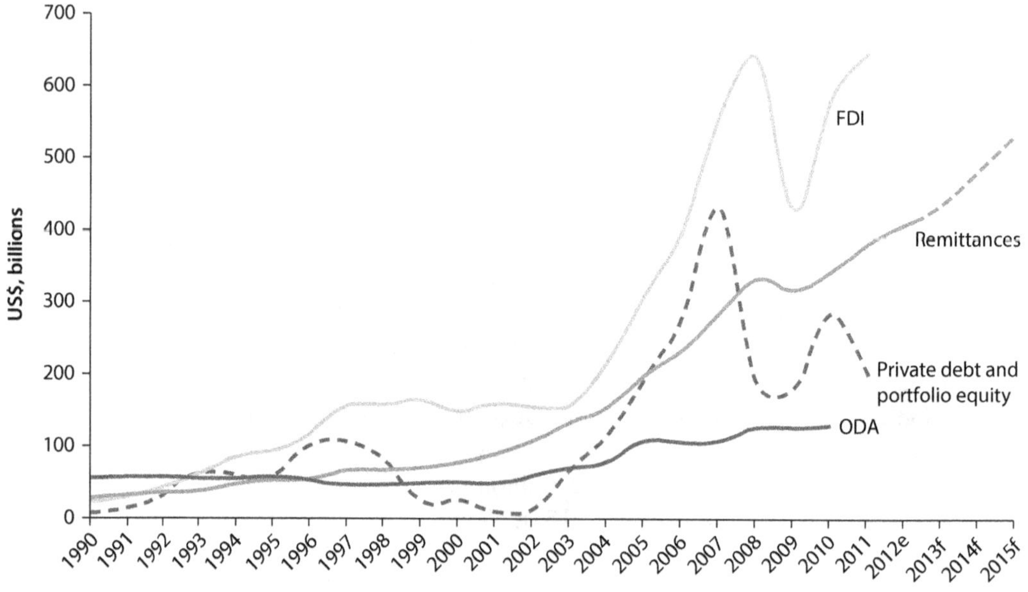

Source: Ratha, Ayana Aga, and Silwal 2012, from World Bank estimates and World Development Indicators Database.
Note: FDI = foreign direct investment; ODA = official development assistance. e = estimated; f = forecast. "Developing countries" refers to low- and middle-income economies as defined in the World Bank's Income Classifications. As of July 1, 2016, low-income economies had 2015 gross national income (GNI) per capita of US$1,025 or less. Middle-income economies had 2015 GNI per capita of US$1,026 to US$12,475.

while for Sri Lanka remittances offset as much as 85 percent of the country's trade deficit in 2010 (Samuel 2016).

Remittances have become a significant source of external funding in the developing world in general, having quadrupled in the past decade, with an apparent opportunity to increase flows to South Asia. According to the International Organization for Migration (IOM), more than 105 million people around the world are migrant workers seeking better opportunities in a foreign country.[11] Global international remittances in 2012 were estimated at US$514 billion, a 10.8 percent increase from 2011, including US$401 billion sent to developing countries (Klapper and Singer 2014).

For South Asian countries, remittances have been a stable source of income, with strong growth potential mostly driven by steady economic opportunities in the Gulf Cooperation Council (GCC) countries.[12] Nepal is among the top recipients of migrant remittances in the world:[13] in 2012, the share of remittances reached 25 percent of its gross domestic product (GDP) and climbed to a staggering 29 percent of GDP by 2013 (table 1.2). In 2012, India was the top recipient in the world in terms of value, with over US$70 billion in receipts; Pakistan was seventh on the list of top earners, with remittance receipts totaling US$14 billion (Ratha, Ayana Aga, and Silwal 2012).

Table 1.2 South Asia Remittance Receipts, by Country, 2009–13
Percentage of GDP

Country	*2009*	*2010*	*2011*	*2012*	*2013*
Afghanistan	1.22	2.08	1.38	1.88	2.65
Bangladesh	10.48	9.79	10.08	10.68	9.24
Bhutan	0.38	0.52	0.57	0.99	0.66
India	3.60	3.13	3.39	3.75	3.73
Maldives	0.23	0.15	0.14	0.15	0.14
Nepal	23.14	21.69	22.37	24.96	28.77
Pakistan	5.19	5.47	5.74	6.24	6.30
Sri Lanka	7.93	8.32	8.71	10.10	9.56

Source: World Development Indicators Database, http://data.worldbank.org/products/wdi.

Impacts of Remittance Methods on Financial Inclusion

Studies on the effects of international remittance transfers on domestic financial systems have confirmed that remittances have a positive impact on financial inclusion and financial sector development (Aggarwal, Demirgüç-Kunt, and Martinez Peria 2006; Anzoategui, Demirgüç-Kunt, and Martinez Peria 2011; Gupta, Pattillo, and Wagh 2009). However, not much evidence pointed to the use of credit markets by the recipients, as the need for credit may have been alleviated as a result of money remitted. Indeed, in South Asian countries with strong remittance flows, the unbanked percentages nevertheless remain high, indicating that poor people still rely principally on cash for daily financial transactions. This is confirmed by the large share of funds tied up as notes and coins in circulation. In South Asia, the ratio of notes and coins is, on average, around 10 percent of GDP—almost twice the rate in developed economies—denoting greater use of cash in the settlement process (Sophastienphong and Kulathunga 2010).

Because of proximity issues, cost, problems with the means of identification, and other factors, senders and receivers of remittances lack safe and affordable ways to save, borrow, and send money. This creates an opportunity for e-money, to the extent that innovative methodologies can solve these problems.

South Asia has the advantage of being the least-cost region to send money to, costing an average of 5.7 percent (of the amount sent) over 2008–15 (World Bank 2015b).[14] Overall, the global average total cost was around 7.7 percent and shows a declining trend. The relatively high costs relate in part to limited competition, Anti-Money Laundering and Combatting Funding of Terrorism (AML/CFT) regulatory restrictions, and the impact of potential penalties on transfer operators that further limit entry into the remittance market (Todoroki et al. 2014).

On the other hand, transaction cost is not the only factor that matters for inclusion. Proximity to a transfer agent matters greatly, since lack of proximity implies travel and opportunity costs such as loss of daily wages. Because transfer agents are largely absent in rural areas, mobile money may offer a solution to these problems.

Although the world has recognized remittance flows as a key contributor to poverty alleviation, little effort has been devoted to improving payment instruments and services through which remittances flow from one country to another or from remitter to receiver. Many remitters are frustrated at the remitting point (capturing end) because of the nonavailability of direct payment instruments and the high cost of transactions. Remittance receivers, for their part, are equally frustrated at the receiving end (disbursement point) because of delays in crediting accounts (transmission delays or bank floats) and high commissions or fees.

Recent innovations by the nonbank money transfer service providers (such as Western Union and MoneyGram) have helped to speed transfer services, but at a cost to remitters and, in some instances, to the remittance receivers as well. Banks have made few attempts to speed up remittance flows at affordable costs in a satisfactory manner. The remittance corridors that have been established have remedied bank-based remittance issues to some extent, although further improvements can be made to payment instruments and services.

It is clear that there is a huge opportunity for South Asian countries to improve financial inclusion—particularly for the poorer segments of the population—by using innovative methods, tools, and applications to harness inward remittance flows. Effective e-money solutions should provide benefits to customers in terms of proximity, safety, reliability, cost, usability, and diversified services that match the needs of the customers, who would, in most cases, otherwise be excluded from the formal financial system.

Notes

1. Findex data from the Global Findex Database, http://datatopics.worldbank.org/financialinclusion/. Findex data are compiled using the Gallup World Poll Survey and measure how adults in 143 economies around the world manage their day-to-day finances and plan for the future. The indicators are constructed using survey data from interviews with more than 150,000 nationally representative and randomly selected adults over the 2014 calendar year (approximately 1,000 people from each country).
2. For the 2014 Global Findex Survey data, see http://datatopics.worldbank.org/financialinclusion/.
3. Global Findex 2014 Survey data, http://datatopics.worldbank.org/financialinclusion/.
4. In 2015 the international extreme poverty and poverty lines were reestimated at US$1.90 and US$3.10 per person per day, respectively, using 2011 purchasing power parity (PPP) exchange rates. The previous extreme poverty and poverty lines had been US$1.25 and US$2.50 per person per day, respectively, using 2005 PPP exchange rates.
5. Global Findex 2014 Survey data, http://datatopics.worldbank.org/financialinclusion/.
6. Although, according to the Honohan (2008) study (table 1.1), 58 percent of South Asian adults do not use or have access to available financial services, the Findex 2011

data showed the rate of financial exclusion to be around 67 percent, and the 2014 Findex exclusion rate is calculated to be around 54 percent (figure 1.1). The difference in data collection methodologies may account for the different estimates. Honohan's study is based on household surveys as well as on data from banks *and* microfinance institutions, many of which may not qualify as "formal" by the Findex definition. The Findex data are culled from a random sampling survey. Arguably, Honohan's study includes more variables and data points that may explain South Asia's microfinance phenomena and the region's higher level of inclusion when compared with Findex 2011 data. The Global Findex 2014 data include not only accounts at formal financial institutions but also formal transaction accounts, such as mobile money.

7. The other South Asian exception, Maldives, is not included in the Global Findex dataset.
8. The number of adults represents around 35 percent of South Asia's total population.
9. The age dependency ratio is a measure showing the number of dependents (ages 0–14 years and over 65 years) in relation to the total working-age population (ages 15–64 years).
10. Age dependency ratios are based on data from the Global Findex Survey 2014 (http://datatopics.worldbank.org/financialinclusion/) and the World Bank's 2015 World Development Indicators dataset (http://data.worldbank.org/products/wdi).
11. For this and other data on international labor migration, see the IOM website: https://www.iom.int/labour-migration.
12. The GCC member states include Bahrain, Kuwait, Oman, Qatar, Saudi Arabia, and the United Arab Emirates.
13. Country remittance data (as a share of GDP) are from the World Bank's 2015 World Development Indicators Database (http://data.worldbank.org/products/wdi).
14. The cost of sending money to East Asia and the Pacific remained substantially stable during the period, at 8.1 percent, while in Latin America and the Caribbean, the cost rose marginally to 6.8 percent (World Bank 2015b). The average for the Middle East and North Africa declined to 8.2 percent, while in Eastern Europe and Central Asia, the average cost was 6 percent including the Russian Federation (7 percent excluding Russia).

Bibliography

Aggarwal, R., A. Demirgüç-Kunt, and M. S. Martinez Peria. 2006. "Do Workers' Remittances Promote Financial Development?" Policy Research Working Paper 3957, World Bank, Washington, DC.

Ahmad, E., and A. Malik. 2009. "Financial Sector Development and Economic Growth: An Empirical Analysis of Developing Countries." *Journal of Economic Cooperation and Development* 30 (1): 17–40.

Anzoategui, D., A. Demirgüç-Kunt, and M. S. Martinez Peria. 2011. "Remittances and Financial Inclusion: Evidence from El Salvador." Working Paper 5839, World Bank, Washington, DC.

Beck, T., A. Demirgüç-Kunt, and R. Levine. 2007. "Finance, Inequality, and Poverty: Cross-Country Evidence." *Journal of Economic Growth* 12 (1): 211–52.

CGAP (Consultative Group to Assist the Poor). 2011. "Global Standard-Setting Bodies and Financial Inclusion for the Poor: Toward Proportionate Standards and Guidance."

White paper for the G20's Global Partnership for Financial Inclusion, CGAP, Washington, DC.

CGD (Commission on Growth and Development). 2008. *The Growth Report: Strategies for Sustained Growth and Inclusive Development*. Washington, DC: CGD and World Bank.

Chen, S., and M. Ravallion. 2010. "China Is Poorer than We Thought, but No Less Successful in the Fight against Poverty." In *Debates on the Measurement of Global Poverty*, edited by S. Anand, P. Segal, and J. Stiglitz, 327–40. Oxford: Oxford University Press.

Claessens, S., and E. Feijen. 2006. "Financial Sector Development and the Millennium Development Goals." Working Paper No. 87, World Bank, Washington, DC.

Creane, S., R. Goyal, M. Mobarak, and S. Randa. 2004. "Financial Sector Development in the Middle East and North Africa." Working Paper No. 04/201, International Monetary Fund, Washington, DC.

Cull, R., A. Demirgüç-Kunt, and J. Morduch, eds. 2013. *Banking the World: Empirical Foundations of Financial Inclusion*. Cambridge, MA: MIT Press.

Cull, R., T. Ehrbeck, and N. Holle. 2014. "Financial Inclusion and Development: Recent Impact Evidence." Focus Note No. 92, Consultative Group to Assist the Poor (CGAP), Washington, DC.

Demirgüç-Kunt, A., and L. Klapper. 2013. "Measuring Financial Inclusion: Explaining Variation in Use of Financial Services across and within Countries." *Brookings Papers on Economic Activity, Economic Studies Program* 46 (1): 279–340.

Demirgüç-Kunt, A., L. Klapper, D. Singer, and P. Van Oudheusden. 2015. "The Global Findex Database 2014: Measuring Financial Inclusion around the World." Policy Research Working Paper 7255, World Bank, Washington, DC.

Grais, W., and A. Kulathunga. 2007. "Capital Structure and Risk in Islamic Financial Services." In *Islamic Finance: The Regulatory Challenge*, edited by S. Archer and R. A. A. Karim, 69–93. Singapore: John Wiley & Sons.

G20 (Group of Twenty). 2009. "G20 Leaders Statement: The Pittsburgh Summit." http://www.g20.utoronto.ca/2009/2009communique0925.html.

Gupta, S., C. A. Pattillo, and S. Wagh. 2009. "Effect of Remittances on Poverty and Financial Development in Sub-Saharan Africa." *World Development* 37 (1): 104–15.

Honohan, P. 2008. "Cross-Country Variation in Household Access to Financial Services." *Journal of Banking & Finance* 32 (11): 2493–500.

Kempson, E., and C. Whyley. 1999a. *Kept Out or Opted Out? Understanding and Combating Financial Exclusion*. Bristol, UK: Policy Press.

———. 1999b. "Understanding and Combating Financial Exclusion." *Insurance Trends* 21: 18–22.

Klapper, L., and D. Singer. 2014. "The Opportunities of Digitizing Payments: How Digitization of Payments, Transfers, and Remittances Contributes to the G20 Goals of Broad-Based Economic Growth, Financial Inclusion, and Women's Economic Empowerment." Report for the G20 Australian Presidency, World Bank, Washington, DC.

Kulathunga, A. 2012. "Can Contemporary Banking Sector Soundness Measures Proxy for Financial Sector Stability and Risk in Developing Countries Pre-financial Crisis and

Mid-financial Crisis?" Doctoral dissertation, The George Washington University, Washington, DC.

Levine, R. 2005. "Finance and Growth: Theory and Evidence." In *Handbook of Economic Growth*, edited by P. Aghion and S. Durlauf, 865–934. Amsterdam: Elsevier Science.

Merton, R. C., and Z. Bodie. 1995. "A Conceptual Framework for Analyzing the Financial Environment." In *The Global Financial System: A Functional Perspective*, edited by D. B. Crane et al., 3–31. Cambridge, MA: Harvard Business School.

Ratha, D., G. Ayana Aga, and A. Silwal. 2012. "Remittances to Developing Countries Will Surpass $400 Billion in 2012." Migration and Development Brief 19, World Bank, Washington, DC.

Samuel, Andrew. 2016. "International Migrant's Day and Sri Lanka." *Sunday Times* (Sri Lanka), December 18.

Sophastienphong, K., and A. Kulathunga. 2010. *Getting Finance in South Asia 2010: Indicators and Analysis of the Commercial Banking Sector*. Washington, DC: World Bank.

Todoroki, E., W. Noor, K. Celik, and A. Kulathunga. 2014. *Making Remittances Work: Balancing Financial Integrity and Inclusion*. Directions in Development Series. Washington, DC: World Bank.

World Bank. 2015a. *Global Monitoring Report 2014/2015: Ending Poverty and Sharing Prosperity*. Washington, DC: World Bank.

———. 2015b. "Remittance Prices Worldwide." Issue No. 14 (June 2015), World Bank, Washington, DC.

———. 2016. *Global Monitoring Report 2015/2016: Development Goals in an Era of Demographic Change*. Washington, DC: World Bank.

CHAPTER 2

Digitizing Financial Inclusion through Innovations

Types of Innovation for Financial Inclusion

South Asia is the birthplace of modern microfinance and today still has some of the world's most sophisticated cash-based financial systems for poor people. It is important to explore how these existing methods of financial inclusion can be augmented through innovative digital financial services (DFSs) or solutions. Technology-based solutions offer tremendous opportunities to transform the landscape of access to financial services for the poor, women, and people living in remote areas. Doing so can spur development and reduce poverty.

Godinho (2010) states that recent advances made by Brazil, China, or India stem to a large extent from building their countries' national innovation systems. However, he also cautions that developing countries will need wise policies as innovation becomes a central component of economic development, as is happening in China, India, and other emerging economies. Appropriate policies will ensure that innovative processes are pro-poor, so that distributional outcomes are socially inclusive.

Developing a technology and innovation strategy requires having an understanding of the different forms of innovation. Henderson and Clark (1990) describe two main forms of innovation:

- *Incremental innovations* exploit the potential of established designs, and often reinforce the dominance of established firms. They improve the functional capabilities of existing technology by means of small-scale improvements in the technology's value by adding attributes such as performance, safety, quality, and cost. Generational or next-generation technology innovations are incremental innovations that lead to the creation of a new but not radically different system.
- *Radical innovations* introduce new concepts that depart significantly from past practices and help create products or processes based on a different set of engineering or scientific principles and often open up entirely new markets and

potential applications. They provide new functional capabilities unavailable in previous versions of the product or service. More specifically related to business, radical innovation has been defined by O'Connor and Ayers (2005) as "the commercialization of new products and technologies that have a strong impact on the market, in terms of offering wholly new benefits, and on the firm, in terms of its ability to create new businesses."

Trends in Financial Service Innovation

Banks typically opt for an incremental launch of innovative products, services, or tools in retail banking. The deregulation and liberalization of the financial sector in the 1970s witnessed the advent of numerous innovations in the form of payment (credit and debit cards); transaction processing (automated teller machines [ATMs], telephone and online banking, and e-commerce for financial assets); saving options (such as investment funds and structured products); loans (automated credit scoring); and risk management techniques (derivatives and securitization). Breakthroughs in information technologies are largely responsible for these new developments that boost productivity, permit a better diversification of risk, and generate economies of scale in internal activities, to mention a few of the benefits (Vives 2010).

Although these innovations brought greater financial inclusion, they also increased a domino-effect contagion, largely due to excessive risk taking, deregulation, and regulatory arbitrage, as evidenced in the global financial crisis of 2007–08. In the wake of the financial fallout and economic downturn, banks have been actively pursuing strategies centered on digital technologies and solutions to provide a holistic service to their clients in an effort to regain their trust, boost flagging customer loyalty, and increase their market share by offering an integrated and seamless customer experience. In both developed and emerging markets, consumers increasingly prefer multichannel banking, with full digital access and more personalized products and services. Interestingly, in these markets, retail banks face serious competition from online-payment specialists and digital merchants such as Square, PayPal, Simple, Google Wallet, Venmo, Amazon, eBay, and the like.

While innovations tend to diffuse across all markets, those same innovations may not generate market development or financial inclusion in developing countries that are at different stages of development and where large segments of society are financially excluded. Moreover, innovations do not always flow in one direction—from more-developed to less-developed markets.

The case studies in this volume highlight innovations in less-developed markets that have not yet been introduced into more-mature markets. The status quo in banking and finance will have to change to accommodate around 2.2 billion poor people in the world (around one-third of the world's population, who are surviving on less than US$2 a day).[1]

What is needed are radical innovations in the way financial services are provided that engender transformational outcomes. Such innovations can create true competitive advantage as well as build new market segments and business opportunities by scaling up. Radical innovations are revolutionary in nature and involve significantly more risk taking, which is why it is rare to see established banking institutions initiating such practices. Breakthrough innovations in financial services could enable large portions of the population to access and use the financial system, thus providing opportunities to work toward better living standards.

Impacts of Recent Innovations on Financial Inclusion

The significance and trends of innovation in retail payment systems and products, along with the impact on financial inclusion from a payment systems perspective, were documented in the 2010 survey conducted in 139 countries by the World Bank's Payment Systems Development Group. The Global Payment Systems Survey observed that there had been a fairly widespread adoption of electronic payment channels for the initiation of payment transactions using innovative retail payment mechanisms (World Bank 2012). In terms of usage, innovative payment products are still used much less frequently than traditional retail payment products. However, they are important for financial inclusion in over 14 percent of the surveyed countries. Although nonbanking entities are playing a significant role in the provision of innovative retail payment products and mechanisms, banks remain a significant player in this field.

Collaboration among various types of entities is widespread, with over one-third of the products involving joint provision of a product or service, almost all of which involved a bank and a telecom company (World Bank 2012). In 60 percent of the cases, customer funds were protected fully. Furthermore, innovative payment products appear to have fairly well-developed pricing models. Merchant payments, utility bill payments, and person-to-person (P2P) transfers were the most common transaction types supported by the innovative payment mechanisms; less than 10 percent of the products supported government-to-persons (G2P) payments.

Most of the innovative products or mechanisms have limited interoperability, with less than 20 percent of the products reported to be fully or partially interoperable (World Bank 2012). Also, traditional clearing and settlement infrastructure is not generally used, and more than 50 percent of the innovative products reported in the survey were settled in the books of the issuer, with only around 24 percent settling in central bank money. Security and fraud risks seem to be getting inadequate attention. Central banks identified themselves as the overseers of around 60 percent of the products; however, 10 percent of the products were subject to collaborative oversight.

In contrast to the detailed transaction data available for traditional retail payment systems and products, the details available for innovative payment products and payment systems are limited. In general, central banks are not overly optimistic about the anticipated impact of innovations on retail payment systems and financial inclusion in their respective jurisdictions. Of the 132 participating central banks, 31 anticipated that the use of electronic payment instruments would increase, 16 foresaw a positive impact on financial inclusion, and 8 expected a positive impact on efficiency (World Bank 2012). Seven central banks anticipated no significant impact on inclusion from ongoing innovations.

E-money and Digital Payments

Though the digital economy is still in its infancy, it is fast taking shape in response to technological advancements and innovations, even in developing countries. Not too long ago, payment systems only dealt with the transfer of money. While electronic payment systems date back to the 1970s, when Western Union introduced the electronic funds transfer (EFT), recent transformative innovations such as mobile applications, e-wallets, near-field communication, and card payments have broadened the array of available electronic payment modalities.

Across the globe, digital payments and e-money products are expanding rapidly, mostly because of their potential to make small-value payments cheaper and safer for both payer and payee. In their article titled "The Future of Global Payments," Bruno, Istace, and Niederkorn (2014) state that the emergence of digital technology is leading to faster and more convenient payments solutions, and a subsequent rise in the expectations of both retail-consumer and commercial clients, thereby expanding financial inclusion. They forecast that payments revenue will grow by 8 percent each year through 2018, at which point annual revenue will reach US$2.3 trillion and account for 43 percent of all banking services revenue, compared with 34 percent in 2009. Although e-money provides a viable means of enhancing financial inclusion, it is not a panacea. Nevertheless, e-money and digital payment systems help address barriers to inclusion and ensure connectivity to mainstream financial services.

What Is E-money?

The World Bank (2012) defines electronic money (e-money) instruments as access mechanisms to prefunded accounts held at banks or nonbank institutions that can be used through the Internet, payment cards, or mobile phones. Such instruments have the potential to further reduce the dependence on paper-based payment instruments by dramatically broadening access to electronic payments for a larger number of consumers, especially unbanked and underbanked consumers.

According to the definition in the "Electronic Money Directive" issued by the European Union, "electronic money" is monetary value as represented by a claim on the issuer that is (a) stored electronically (including magnetically); (b) issued on receipt of funds of an amount not less in value than the monetary value issued; and (c) accepted as a means of payment by undertakings other than those of the issuer (ECB 2009).

It is not necessary for e-money to be associated with a bank in making the payment transaction, because e-money acts as a prepaid bearer instrument.

Table 2.1 Differences between Electronic Money and Virtual Currency Schemes

Aspect	*Electronic money schemes*	*Virtual currency schemes*
Money format	Digital	Digital
Unit of account	Traditional currency (euro, U.S. dollars, pounds, etc.) that have legal tender status	Invented currency (Linden Dollars,[a] bitcoins, and other types) without legal tender status
Acceptance	By undertakings other than the issuer	Usually within a specific virtual community
Legal status	Regulated	Unregulated
Issuer	Legally established electronic money institution	Nonfinancial private company
Supply of money	Fixed	Not fixed (depends on issuer's decisions)
Possibility of redeeming funds	Guaranteed (and at par value)	Not guaranteed
Supervision	Yes	No
Type(s) of risk	Mainly operational	Legal, credit, liquidity, and operational

Source: ECB 2012.
a. Linden Dollars are a currency used in the virtual world Second Life (SL), whereby SL users buy from and sell to one another using the Linden, which is exchangeable for U.S. dollars or other currencies on market-based currency exchanges.

Virtual currency schemes such as Bitcoin are a form of e-money; however, unlike virtual currency, the link between e-money and the fiat currency against which the e-money is issued remains intact, because funds are expressed in units of that currency such as U.S. dollars or euros (table 2.1).

It should also be noted that there is an ongoing debate on whether e-money can be considered "money." The principal objections raised, as discussed by Geva and Kianieff (2005), are the following: First, e-money does not provide a distinct unit of account. Second, payment with e-money may not be anonymous, because the third-party obligor may keep a record of each transfer. Third, e-money may be created other than by withdrawal from a reservable deposit with a commercial bank, so as to undermine the central bank's hold on monetary policy. Fourth, e-money cannot constitute legal tender. However, these conditions can be addressed (though not entirely eliminated) if e-money is universally accepted and if the economy becomes truly digital.

The impact of developing e-money on a country's monetary policy is a fundamental concern. Essentially, monetary policy is aimed at preserving price stability; to do that, the monetary authority manages the money supply through various policy tools. If e-money is created by converting cash or deposit accounts, there won't be new money creation, hence no additional impact on monetary policy. Consequently, digital financial applications typically ensure that the total value of e-money is mirrored in a bank account in the banking system. Although the bank account can earn interest on the balance, e-money in the digital payment system cannot.

Digital Payment Systems

The World Bank (2012) analysis of the evolution of retail payments over the past five to six decades shows that

- Successful adoption of advances in technology have played a key role in the development of new channels for payment initiation, improved authentication, and efficient processing;
- The development of new payment needs (transit payments, Internet auction sites, and social networking sites) have also led to the creation of new payment mechanisms; and
- Payment infrastructure created for one payment product has been successfully leveraged for other payment products, such as using an automated clearinghouse (ACH) for online banking–enabled payments or the successful leveraging of infrastructure created for credit cards by debit cards.

Payment systems enable people and businesses to pay bills, buy goods and services, remit money, and collect dues. Moreover, they help governments and regulators to pay salaries, pensions, and other remunerations, disburse social grants and other benefits; and collect taxes, fees, and dues from the general public. Transformative innovations in the payment systems offer alternative digital channels, such as real-time payments using mobile phones and smartphones, cards, and Internet, thereby enabling faster, more secure, and more efficient services at lower cost. While some of these options may not work for the financially excluded, others, such as ubiquitous access to mobile phones, can enable efficient and low-cost payment alternatives to traditional methods.

Tarazi (2011) highlighted the cost implications of branch versus branchless banking, demonstrating that significant savings can be achieved in moving toward branchless banking, which would positively affect the value propositions involved in offering services to poor people (figure 2.1). Often, digitizing payments is the first step toward becoming a cash-lite society. Many countries may not go beyond this point, but even so, the impact of this measure alone on financial inclusion would be significant.

Toward a Cash-Lite Society

Although financial inclusion is increasingly becoming a national strategy in many countries, embracing digital payments and moving toward a cash-lite society rarely is seen as a national development priority. Lack of visible and tangible results in the short run may be the reason. However, growing evidence on the benefits offered by electronic payments over cash in terms of safety, convenience, least cost, fighting crime and corruption, and advancing financial inclusion suggests that a strategically driven and coordinated approach managed by regulators and policy makers, along with bank and nonbank operators, is needed to unlock the potential of DFS technology in developing countries (box 2.1).

Figure 2.1 Sample Relative Costs of Payment System Infrastructure, from Bank Branches to Mobile Phone

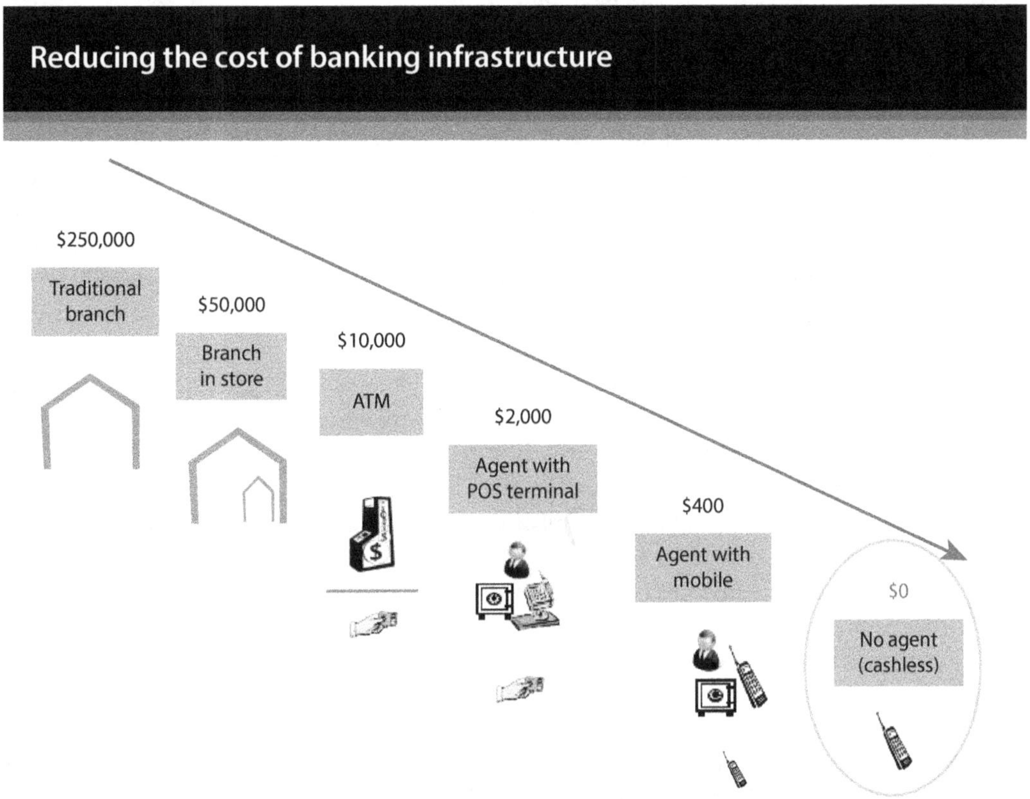

Source: Tarazi 2011. © Consultative Group to Assist the Poor (CGAP). Reproduced, with permission, from CGAP. Further permission required for reuse.
Note: ATM = automated teller machine; POS = point of sale. Dollar amounts shown indicate the approximate capital cost of each technology.

Box 2.1 Cash versus Electronic Payments

Today, around 85 percent of all retail payment transactions are done with cash, which equates to 60 percent of retail transaction value. Although much of the world's population has access to many options for making payments other than cash, it continues to persist as a major means of exchange. Cash takes time to get at, is riskier to carry, and by most estimates, costs society as much as 1.5 percent of gross domestic product (GDP).

Electronic payments, on the other hand, have been proven to boost economic growth, while advancing financial inclusion. It is for these reasons that countries around the world are working to make their payment systems less dependent on cash.

Source: Thomas, Jain, and Angus 2013.

Even today, many central banks and policy makers in the bankcentric developing countries consider payments as part of information and communication technology (ICT) and information technology (IT) and continue to ignore the potential of mobile-based payment systems. This is primarily because of ignorance and also a vested interest in traditional financial service providers—that is, banks and approved financial institutions.

On the contrary, many advanced-country central banks have appointed national payment councils (NPCs) to educate all stakeholders involved in payment system development for both large-value and retail payments. Although a tremendous amount of progress has been achieved in advanced markets such as Australia, Canada, and the United Kingdom, little or no achievement is recorded in developing- and emerging-country NPCs.

Some central banks have become "onlookers" as market practitioners, and financial market infrastructure (FMI) providers have overtaken them by keeping up with emerging trends in technology. In this context, it is imperative to enhance the capacities and knowledge of senior policy makers and regulators as to the wide-ranging benefits of payment system development when it comes to achieving broader social and economic advantages—including those related to effective monetary management, the establishment of financial system stability, poverty reduction, and the promotion of financial inclusion and access to finance.

In most developing countries, adopting these systems can be a revolutionary experience by providing access to a previously unbanked population. Digitizing payments can lead to tangible efficiencies in the form of huge cost savings and increased efficiency for payers, as in the following examples:

- *In Mexico,* by digitizing and centralizing its payments, the Mexican government is saving an estimated US$1.27 billion per year—or 3.3 percent of its total expenditure—on wages, pensions, and social transfers. In 2012, 97 percent of pension payments were made by electronic transfers (Babatz 2013).
- *In Kenya,* mobile payments made through M-Pesa played a central role in improving financial inclusion—increasing the share of adults having access to financial services from 18.9 percent in 2006 to 66.7 percent in 2013 (FSD Kenya and CBK 2013) and clearing K Sh 192.6 billion (US$2.09 billion) worth of mobile transactions in March 2014.
- *In Brazil,* when the government disbursed Bolsa Família grants through electronic benefit cards, administrative costs were reduced from 14.7 percent to 2.6 percent of the total grant value (Pickens, Porteous, and Rotman 2009).

Transitioning toward a cash-lite society starts with digitizing the payments channel. Bankable Frontier Associates (BFA 2012) identifies four stages on the path from a "cash-heavy" society at one end (in which cash is by far the predominant payment instrument) toward a "cash-lite" society at the other (in which cash is no longer the most common means of payment), as follows (figure 2.2):

1. In a cash-heavy society, paper-based instruments (such as cash or perhaps checks) are the main ones in use.
2. The first shift happens when bulk payers in an economy—such as government, large employers, or development aid distributors—decide to pay electronically.
3. The second shift takes place as opportunities grow for recipients to spend or transfer money electronically.
4. Finally, a cash-lite society has been achieved when even the majority of small payments—usually transactions between people and merchants (person-to-business [P2B]) for everyday items like groceries—also become electronic.

Figure 2.2 Stages and Shifts from a Cash-Heavy to a Cash-Lite Society

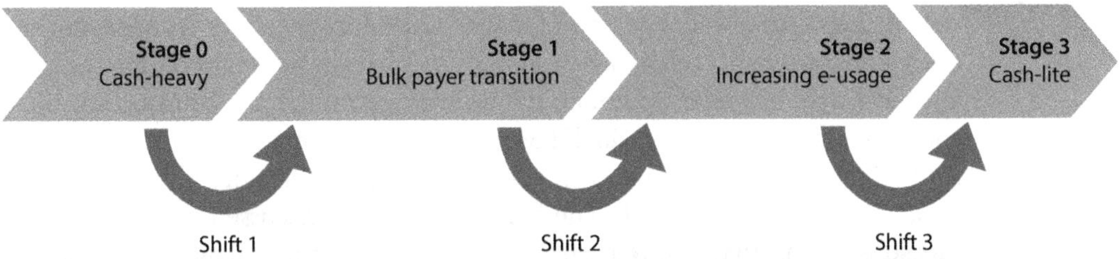

	Stage 0 Cash-heavy	Stage 1 Bulk payer transition	Stage 2 Increasing e-usage	Stage 3 Cash-lite
Flow of electronic payments		Few to many	Many to few	Many to many
Main payment instruments in use	Mainly paper (typically cash; maybe some checks)	Mixture: paper and electronic (cards used at ATMs, some online banking)	Mainly electronic (mobile used for bill payments and remittances)	Almost all electronic (use of mobile and/or card at point of sale through inter-connected switches)
What is needed to shift to this level?		Sufficient cash-out points; B2P and G2P shifts	Ability of business and consumers to make cheap electronic payments via computer, standing order, ATM (P2P, P2B)	Pervasive acceptance of electronic payments at POS and mobile phone, compelling financial products
Examples	Haiti, Niger	Colombia	Kenya	Canada, northern Europe, United States

Source: BFA 2012. © Bankable Frontier Associates (BFA). Reproduced, with permission, from BFA. Further permission required for reuse.
Note: ATM = automated teller machine; B2P = business-to-person; G2P = government-to-person; POS = point of sale; P2B = person-to-business; P2P = person-to-person.

The BFA (2012) study recognizes that these shifts may not always be linear; however, each shift can bring significant benefits in terms of cost, reliability, safety, proximity, and even access to new markets and opportunities for users.

The earlier examples from Brazil, Kenya, and Mexico provide evidence to amply demonstrate that the use of innovative e-money and digital payment solutions has brought previously excluded people into the formal financial system. These developments are encouraging for cash-based South Asian countries, where only 46 percent of adults are financially included.[2] Moving away from cash, therefore, would continue to open access to the formal financial system for many of South Asia's unbanked.

Risks in Digital Finance

As the detailed explanation in Part II of this volume (chapters 4–6) suggests, technology alone does not bring truly transformational outcomes, and technology is certainly not a "silver bullet." It is the human element—vision, long-term commitment, cooperation between public and private stakeholders, and trust—at each stakeholder level that provides the leadership necessary to confront and overcome barriers to achieving greater levels of financial inclusion for the poor in a sustainable manner. Although the technology, model, transmission channels, and products may be new in DFSs, traditional risk management methods of identifying risks and of quantifying and assessing the likelihood of occurrence remain true for DFSs as well. The following risks (as discussed in more detail in the subsections that follow) can derail the successful scaling-up and use of e-money at all stakeholder levels:

- *Political economy risks* stem from vested interests that resist the introduction of disruptive technologies, as well as legal and regulatory risks arising from lack of governance codes, payments laws, Anti-Money Laundering and Combatting Funding of Terrorism (AML/CFT) laws, other laws and regulations.
- *Cybersecurity risks* may result from the failure to enable institutions with the capacity to ensure the efficient functioning of systems and protect against cyberthreats.
- *Risks to the payments system* may compromise system functions if the service providers do not adequately understand the digital customer base or the infrastructure requirements.
- *Principal-agent risks* must be addressed to minimize revenue losses, fines and other reprimands by the regulators, fraudulent activities and corruption, and loss of reputation.
- *Risks to customers* result not only from inadequate information and understanding of e-money but also from lack of consumer protection and adequate redress mechanisms, identity theft, and liquidity-related issues, among others.

Political Economy Risks

Digital finance is a "disruptive technology," especially when it is operated by nonbanks such as technology companies or mobile network operators (MNOs). Existing regulated financial institutions represent vested interests that may oppose the introduction of nonbank competitors, whose lower-cost business model is seen as unfair competition and a threat to existing bank-based operations and potential profits.[3]

Although the economics of e-money at the macro level are driven by supply and demand and the fixed cost of providing infrastructure, political economy considerations shape the issue of regulating e-payments. Given the influence of special interests and their collective powers, often the strength and leadership of the regulator alone determines the success or failure of perfectly legitimate e-money initiatives. This aspect is highlighted in the case studies.

The regulator's role in setting and achieving financial inclusion objectives includes leveling the playing field by introducing regulations that enable entry and competition by these "disruptors" while putting in place flexible and effective risk management frameworks to protect consumer funds. This may require adjusting know-your-customer (KYC) and registration requirements that limit uptake and use. The outcome substantially affects the business risk to the digital financial operator because the large, long-term nature of the investment in infrastructure and customer touchpoints requires widespread customer use for the venture to be cost-effective and eventually profitable. In some cases, the unwillingness of the regulator to challenge vested interests—by facilitating nonbank competitors' entry—partly explains the slow pace of growth of successful e-money deployments.

Cybersecurity Risks

In promoting financial inclusion through digital payment instruments and channels, policy makers and regulators should also pay attention to external disruptive elements such as cybercrimes that can compromise databases and the functioning of technology-based products and services. Although policy makers have urged market participants to work together to combat cybercrimes, service providers in most emerging and developing markets have not taken such calls seriously enough. Nor have they consented to coordinate or cooperate with peers in reducing costs involved in establishing preventive measures, including legal and regulatory measures.

In general, high-value payment systems, such as real-time gross settlement systems, are closely monitored, and service providers observe best practice and work within rigorous regulatory frameworks. As such, high-value payment systems are relatively safer than retail payment services, which consist of an array of payment instruments and channels, mostly digital. Retail service providers include nonbank institutions, whose commitment to prudent payment services can be questionable, given the scale and use of ad hoc infrastructure platforms. Whereas potential damage to a high-value payment system can be far more severe, given the high volume and value as well as the criticality of such transactions, retail payment services also require prudence and effectiveness in surveillance and monitoring.

The Bank for International Settlements Committee on Payments and Market Infrastructures (BIS/CPMI) 24 guidelines[4] are important to ensure that digital services and devices are safe and do not exclude the poor, as these guidelines have been prepared to capture operations of all financial market infrastructure providers and participants. In the local context, the applicability of some of the existing legal and regulatory frameworks may not be adequate to cover digital payment services. Hence, policy makers and regulators may need technical advice and support from international organizations in dealing with external disruptions such as cybercrime.

Although relatively small in value, a cyberthreat against retail and micro-level digital finance service providers can have serious consequences for the poor and vulnerable. The micro-level service providers can be wiped out, while the clientele can lose their lifetime savings and wealth (if any), even under a mild form of cyberattack. Unfortunately, because of heavy competition as well as their small scale and lower profitability, many service providers in developing countries do not appear sufficiently concerned about the potential impacts of cyberattacks to observe international best practice cyber protocols and security guidelines intended to cushion the impact of cyberthreats. In this regard, efforts are needed by both international and national bodies that deal with cyberthreats to coordinate and promote the observance of international protocols and best practice by DFS providers.

Risks to the Payments System

The essence of a payments system is to offer transactional services with the minimum amount of risk. Digital payments as narrowly discussed in this study do not introduce new types of risks to the payment system, and digital payment systems are generally not considered to be systemically important (with the exception of Kenya's M-Pesa and the South African Social Security Agency [SASSA] social grants system). It is important to note that large-value payments are typically managed by central banks, which have the ability to address impending risks online on a real-time basis. Other retail payment systems typically handle transactions between prudentially regulated financial institutions that have their own risk management frameworks, already vetted by the central banks.

In contrast, new digital technologies often result in large numbers of low-value transactions, scattered across a country and bringing in a vast number of players (both payment system providers and their agent networks) with little if any prudential training, risk management abilities, or understanding of the types of risks and requirements of handling financial transactions. It is for this reason that most regulators are often hesitant to bring nonbank operators into the payment arena. In this regard, there may be a case for establishing an NPC to play a role in ensuring that such payments risks are minimized. Nevertheless, the NPCs often are seen to pay scant attention to the retail payment systems and focus mostly on the large-value systems. The case studies reflect the importance of developing e-money, as defined in this report, as an integral part of a country's national payments framework that is covered by the payments law and enabling provisions.

Principal-Agent Risks

The principal-agent risk or problem arises because the MNO or DFS provider must depend upon a network of agents to provide on-the-ground services directly to the customer. Because of the large geographical distribution of agents, network management often is employed to monitor and manage agent performance and ensure that agent incentives are aligned with those of the principal, the DFS provider. Agents and agent networks introduce new operational, financial crime, and consumer risks, many of which are due to the physical distance between agents and the provider or the agent network manager and the resulting challenges to effective training and oversight. Operational risks include fraud, agent error, poor cash management by the agent, and poor data handling. In addition to the financial crime risks of fraud and theft (including data theft), agents may fail to comply with AML/CFT rules regarding customer due diligence, handling records, and reporting suspicious transactions. Agents may also take actions that reduce transparency (for example, on pricing, terms, and recourse), engage in abusive treatment of customers (including overcharging), or fail to handle customer data confidentially (McKee, Kaffenberger, and Zimmerman 2015).

Risks to the Customer

At the consumer level, evidence from research conducted in 16 countries by McKee, Kaffenberger, and Zimmerman (2015) reveals information on customers' perceptions and experiences with risks and the ways in which these risks harm their trust, uptake, and use of the services. Although customers highly value and benefit from many basic DFSs, many users are not only new to both formal finance and technology but also live precarious financial lives that allow little room for error.

Also revealed is strong evidence that providers are actually not realizing the full potential of growing numbers of DFS deployments. In many cases this is because customers experience challenges or problems that erode their trust and therefore limit their use of DFSs—often to over-the-counter (OTC) transactions, which do not require an account. When nonusers (potential customers) observe friends and family struggling with digital platforms, they frequently conclude that the services are simply too risky. The report identified seven key risk areas for customers of DFSs (box 2.2).

Such risks can translate into loss of savings, indebtedness, loss of identification, and loss of faith in the financial system, leading to reexclusion from the financial system. Practical steps undertaken to protect customers from the risks of digital finance and to facilitate consumer trust, uptake, and use of e-money are addressed in the case studies at different stakeholder levels.

Box 2.2 Doing Digital Finance Right: The Case for Stronger Customer Risk Mitigation

Research in 16 countries concerning customers' perceptions of and experiences with digital financial services (DFSs) revealed seven perceived risk areas:

- *Inability to transact due to network or service downtime.* This most commonly cited risk area can lead to risky customer behaviors such as leaving cash with an agent to conduct a transaction later when the network is back up. It also presents challenges if customers need money urgently and cannot cash out until the network resumes.
- *Insufficient agent liquidity.* This is the second most common risk-related area that commonly prevents customers from transacting and accessing their money. This problem particularly plagues bulk payment recipients (such as G2P [government-to-persons] recipients): because many receive their transfers all on the same day and want to cash out immediately, agents struggle to meet liquidity demands. The recipients are often among the poorest in a country, and the extra fees and the delay in receiving their benefits can seriously affect their ability to meet basic needs.
- *Complex and confusing user interfaces.* This customer risk is exemplified by the difficult-to-resolve experience of a user who sends money to a wrong number, which often also results in financial loss. Difficulties with the menu also cause many customers to seek assistance conducting transactions, requiring them to share private information (such as their personal identification number [PIN]) with an agent, family member or friend—exposing them to potential fraud by the person providing help.
- *Inadequate provider recourse.* Complaints and dispute resolution options are unclear, causing the customer to lose time, money, and mobile-phone airtime to either travel to customer care centers or wait on hold for call center staff who may or may not be able to solve the problem.
- *Nontransparency of fees and other terms.* This issue prevents customers from fully understanding the details of services and leaves them vulnerable to agent misconduct and price fraud.
- *Fraud perpetrated on customers.* Customers can experience fraud at the hands of provider employees or external fraudsters who use "social engineering" scams such as phony promotions to obtain money or information from unsuspecting customers.[a] Agents can also perpetuate fraud by charging unauthorized fees.
- *Inadequate privacy and protection of customers' personal data.* Disclosure of data-handling practices is often weak, with details available only on a website to which few consumers have access and in "legalese" that is difficult to understand.

Source: McKee, Kaffenberger, and Zimmerman 2015.
a. "Social engineering scams" refers to scams whereby a hacker tricks the target themselves, or perhaps an unwitting customer service agent, into revealing information that gives the hacker access to victims' passwords or accounts.

Notes

1. Data from the Poverty Topic Overview, World Bank, http://www.worldbank.org/en/topic/poverty/overview. US$2 a day represents the median poverty line of developing countries in 2005 purchasing power parity (PPP) terms (Chen and Ravallion 2010).
2. South Asian financial inclusion data from the Global Findex Survey 2014, http://datatopics.worldbank.org/financialinclusion/.
3. This opposition persists despite the fact that the target clientele for digital payments are generally the unbanked and the poor, whose market needs, multitude of small transaction values, and remote locations are extremely costly and not at all interesting to traditional financial institutions.
4. In June 2014, the Committee on Payment and Settlement Systems (CPSS) was renamed as the Committee on Payments and Market Infrastructures (CPMI) under a new charter. Hence the BIS/CPSS 24 guidelines issued in 2012 (BIS and IOSCO 2012) were renamed the BIS/CPMI 24 guidelines.

Bibliography

Babatz, G. 2013. "Sustained Effort, Saving Billions: Lessons from the Mexican Government's Shift to Electronic Payments." Evidence Paper, Better Than Cash Alliance, New York.

BFA (Bankable Frontier Associates). 2012. "The Journey toward 'Cash Lite': Addressing Poverty, Saving Money, and Increasing Transparency by Accelerating the Shift to Electronic Payments." BFA study for the Better Than Cash Alliance, Somerville, MA.

BIS and IOSCO (Bank for International Settlements and International Organization of Securities Commissions). 2012. *Principles for Financial Market Infrastructures*. Basel: BIS; Madrid: IOSCO.

Bruno, P., F. Istace, and M. Niederkorn. 2014. "The Future of Global Payments." Excerpt from "Global Payments 2014: A Return to Sustainable Growth Brings New Challenges," a report of the Financial Services Practice, McKinsey & Company, New York.

Chen, S., and M. Ravallion. 2010. "China Is Poorer than We Thought, but No Less Successful in the Fight against Poverty." In *Debates on the Measurement of Global Poverty*, edited by S. Anand, P. Segal, and J. Stiglitz, 327–40. Oxford: Oxford University Press.

ECB (European Central Bank). 2009. "Electronic Money Directive (2009/110/EC)." Directive 2009/110/EC of the European Parliament and of the Council of 16 September 2009. *Official Journal of the European Union* L 267: 7–17.

———. 2012. *Virtual Currency Schemes*. Frankfurt: ECB.

FSD Kenya and CBK (Financial Sector Deepening Kenya and Central Bank of Kenya). 2013. "FinAccess National Survey 2013: Profiling Developments in Financial Access and Usage in Kenya." Survey results report, FSD Kenya and CBK, Nairobi.

Geva, B., and M. Kianieff. 2005. "Reimagining E-money: Its Conceptual Unity with Other Retail Payment Systems." In *Current Developments in Monetary and Financial Law, Volume 3*, 669–705. Washington, DC: International Monetary Fund.

Godinho, M. Mira. 2010. "Economic Development Revisited: How Has Innovation Contributed towards Easing Poverty?" In *INNOVATION: Perspectives for the 21st Century*, 269–85. Madrid: BBVA.

Henderson, R. M., and K. B. Clark. 1990. "Architectural Innovation: The Reconfiguration of Existing Product Technologies and the Failure of Established Firms." *Administrative Science Quarterly* 35 (1): 9–30.

McKee, K., M. Kaffenberger, and J. M. Zimmerman. 2015. "Doing Digital Finance Right: The Case for Stronger Mitigation on Customer Risks." Focus Note 103, Consultative Group to Assist the Poor (CGAP), Washington, DC.

O'Connor, G. C., and A. D. Ayers. 2005. "Building a Radical Innovation Competency." *Research-Technology Management* 48 (1): 23–31.

Pickens, M., D. Porteous, and S. Rotman. 2009. "Banking the Poor via G2P Payments." Focus Note 58, Consultative Group to Assist the Poor (CGAP), Washington, DC.

Tarazi, M. 2011. "Branchless Banking and Financial Inclusion." PowerPoint presentation to the Consultative Group to Assist the Poor (CGAP), Washington, DC, June 2.

Thomas, H., A. Jain, and M. Angus. 2013. "MasterCard Advisors' Cashless Journey: The Global Journey from Cash to Cashless." White paper on the Cashless Journey Study, MasterCard Advisors, Purchase, NY.

Vives, X. 2010. "The Financial Industry and the Crisis: The Role of Innovation." In *INNOVATION: Perspectives for the 21st Century*, 321–29. Madrid: BBVA.

World Bank. 2012. "Innovations in Retail Payments Worldwide: A Snapshot. Outcomes of the Global Survey on Innovations in Retail Payment Instruments and Methods." Report of the Global Payment Systems Survey 2010, World Bank, Washington, DC.

CHAPTER 3

Stakeholders in Digital Financial Inclusion

Introduction

In examining how financial inclusion can be advanced through the use of digital channels, four levels of stakeholders that are essential to financial inclusion pathways can be identified:

- *Macro:* policy makers, regulators, and donors
- *Meso:* organizations that support financial services providers
- *Micro:* digital financial services (DFSs) providers
- *Customer*

Digital innovations that can influence financial inclusion through these pathways include e-money, mobile-phone access to e-wallets or bank accounts, card systems, near-field communication (NFC)–enabled systems, agent and correspondent financial services, remittances, and government-to-persons (G2P) and person-to-government (P2G) payments through partially or fully electronic means. To determine which innovations in the digital space can enhance financial inclusion, it is important to identify the ways in which customers—as well as micro-, meso-, and macro-level actors—can be influenced and leveraged to achieve greater financial inclusion using innovative digital means and pathways. To that end, each level has a role to play in promoting digital financial inclusion:

- *Policy makers* must provide leadership and vision in setting the policy agenda, and *regulators* should develop an enabling, nonprohibitive regulatory environment that may be iteratively layered as the market matures.
- *Enabling institutions* should develop a rich ecosystem and infrastructure in which digital finance can grow.

- *DFS providers* should develop and operate various digital solutions and options that create, preserve, and enhance positive value propositions for the customers who are predominantly from the base of the pyramid.
- *Customer* preferences should be identified, and customers must be given the opportunity to acquire financial education to better understand how they can benefit from financial inclusion and use digital financial services to manage their needs as well as access other economic opportunities in a much more cost-effective manner.

Macro-Level Stakeholders: Policy Makers, Regulators, and Donors

Macro-level stakeholders can be defined as national policy makers, donors, funders, national regulatory and supervisory bodies, and even international knowledge exchange and advocacy bodies. The goals of most of those listed are either to perform or direct micro- or meso-level interventions, or to influence national governmental players. From the donors' perspective, macro-level interventions are the second-most funded interventions after micro-level. Indeed, many multilateral donors have separate departments to support public (macro) and private (micro and meso) sector development.

Governments as policy makers are the key players in driving the national digital financial agenda, and they oversee coordinated policy actions. In some developing countries today, it is heartening to see government agencies themselves developing policies, drafting legislation, sharing knowledge among themselves (in organic South-South knowledge exchanges), and achieving such levels of innovation that donors can be seen trying to keep up. In such instances, governments are seen to be managing the transition to DFSs not only by mandating but also by consciously supporting the transition.

Regulators are macro-level stakeholders whose responses to financial innovation need careful attention. From the point of view of financial systemwide stability, the apprehensions of central banks and regulators should be appreciated but should not inhibit the much-needed promotion of financial inclusion. It is important for central banks and regulators to permit and even facilitate technological innovations insofar as they promote financial inclusion and poverty reduction, while ensuring that they do not outstrip or circumvent prudential regulations to maintain the safety of the financial system and customers' funds. An enabling regulator should focus on policies and strategies that target financial inclusion and broaden financial access through the entry of new players, while regulating risks associated with lower-cost intermediaries entering the financial space.

However, in a highly dynamic, innovative digital space, imperfect knowledge about digital products and systems leaves regulators permanently in catch-up mode. As digital systems evolve and new players enter the payment space, new kinds of risks emerge, especially outside the banking system. Concerns about Anti-Money Laundering and Combatting Funding of Terrorism (AML/CFT) and consumer protection issues have to be addressed quickly and efficiently.

Hence, regulators have to be proactive while avoiding regulatory overkill that might stifle financial innovations that can expand financial inclusion of the unbanked and underbanked masses.

Hence, central banks and regulators should step up their supervision and oversight capabilities rather than refuse approvals or use delaying tactics to prevent nonbanks—mobile network operators (MNOs), in particular—from entering into payment spaces. Furthermore, they should offer technology-based payment solutions to deal with basic payment transactions (such as cash-in/cash-out in domestic remittances), bill-pay facilities, and similar transaction services as an important stepping-stone toward entering the formal financial system.

Meso-Level Stakeholders: Enabling Institutions

Meso-level stakeholders do not provide direct financial services but strengthen and support mainly micro-level players. They typically provide information, support services, and infrastructure and include the following:

- *Technical services providers* for consumer and financial institution training, consulting, and product support, potentially for each product type
- *Industry research organizations* such as research institutes and universities, consulting firms, and rating agencies
- *Policy advocates, industry associations, and networks* (local, regional, and international)
- *Technology service and payment service providers* for hardware, software, and information technology (IT) technical support for payments systems; automated teller machines (ATMs); core banking software; and mobile banking
- *MNOs* that provide services on behalf of micro-level players
- *Agents,* such as agent networks of MNOs, because although they are the customer touchpoints for mobile money, they act on behalf of MNOs in a supporting role

Not all of these players participate in developing digital finance systems in each country. Their involvements will vary with the level of development of the systems, donor activities, regulatory and policy structures, and so on. They support private sector operators in setting up and managing new systems, as well as strengthen the existing micro-level actors to be more financially sustainable and sophisticated (and thereby offer better service to their clients). They also influence the micro-level players to extend their client base and provide knowledge, training, and information on national and international best practice. Meso-level stakeholder support and participation in the development process will also reduce market distortions and enhance the efficiency and effectiveness of the micro-level stakeholders. In addition, they are well positioned to provide financial education services to the consumers and can play an advisory role to the macro-level stakeholders as well.

The 2012 Bank for International Settlements Committee on Payments and Market Infrastructures (BIS/CPMI) 24 guidelines for financial management information systems (FMISs) recognize the critical importance of all service providers in payment system development and the need for compliance with such guidelines for safety, efficiency, reliability, and affordability of payment services (BIS and IOSCO 2012). In a way, these guidelines bring market discipline—a lacuna in the payment landscape for some time—to all payment system providers.

Although meso-level service providers and stakeholders are also expected to observe standards and market ethics, the regulators are expected to monitor observance through effective surveillance systems. Some countries in Asia (Malaysia and Singapore) have already complied with many of the CPMI guidelines—as seen from the endorsements received from the International Monetary Fund's and World Bank's Financial Sector Assessment Program evaluation teams—while other countries seem to lag behind. It is important that national payment councils in other countries take the initiative to educate meso-level service providers on the importance of observing these best practices to develop healthy payment systems (for retail payments, in particular) that offer numerous benefits in promoting financial inclusion and access to finance.

Micro-Level Stakeholders: Institutions Offering Digital Solutions

Micro-level stakeholders are entities that provide DFSs to customers. They include banks and other licensed financial institutions; nonbank financial institutions that are authorized to operate mobile money solutions (typically MNOs); authorized e-money issuers; and other payment service providers.

It is important to have a national agenda on digital financial inclusion so that micro-level service providers are able to play various roles: fill gaps in the market; provide services to underserved areas via relaxed regulatory hurdles, moral suasion, quotas, or subsidies; offer additional product offerings through strategic partnerships and product developments; and strengthen their institutional capacity to increase institutional sustainability and sophistication. The goal is broader outreach, better-quality products, improved customer interactions, and sharing of national and international best practices to facilitate institutional self-improvement.

Often, digital financial inclusion strategies stall because some of these micro-institution types that have touchpoints with financially excluded people are overlooked. E-wallets and mobile-phone access are two frequently neglected channels. Micro-level service providers should be persuaded to use less cash and paper-based instruments, and medium-term targets should be set for the purpose.

Customer-Level Stakeholders: Users

Customers are the people and institutions that use the DFS and payment services. Sometimes, when financial products or solutions fail, it is because they are supply-driven—designed to deliver industry perceptions of what customers need rather than serving an unmet need expressed by actual customers. Focusing on customer convenience and efficiency will guarantee a successful digital financial deployment. It is, therefore, important for stakeholders to understand their customers' true perceptions, needs, and beliefs.

Although the pathways to financial inclusion start with the first customer touchpoint, which is usually based on a specific need, there is no guarantee that just because a customer had previously used a particular financial service, he or she will then go on to learn about or obtain access to others. Use and trust in the reliability, convenience, and safety of the service needs to be promoted to maximize its chances of success. It is important to engage community-level outreach, especially in rural communities, and to provide financial education programs. It is also important to understand that, in many countries, first-time users commonly prefer to use over-the-counter transactions.

Service providers must understand the needs of these populations, building trust and facilitating their familiarity with digital solutions in order to convert them into account holders. Although customers may gradually give up their predilection for cash- and paper-based payment instruments, effective incentives—including price incentives—may be provided to accelerate progress toward the goal of paperless payments in the interest of financial inclusion.

Bibliography

BIS and IOSCO (Bank for International Settlements and International Organization of Securities Commissions). 2012. *Principles for Financial Market Infrastructures*. Basel: BIS; Madrid: IOSCO.

PART II

Critical Enablers That Are Game Changers in Successful E-money Deployments

Transitioning to digital payments will require well-supported evidence that doing so is in the interest of all participants—businesses, governments, financial institutions, merchants and consumers. Investment will be required to update systems to replace paper payments and facilitate electronic processing of invoices. None of this will happen without first presenting a tangible case for future success.

—Canadian Task Force for the Payments System Review,
"Going Digital: Transitioning to Digital Payments"

The main reasons why people lack access to formal financial systems seem to be similar across the globe. Those reasons include poverty, lack of proximity, cost, not having documentation or identification, lack of financial literacy, low acceptability, risk aversion, and low levels of trust. Cultural factors also play an important role. Banks and most formal financial institutions are unwelcoming to the poor, whose numerous and tiny transactions would be unacceptably costly for most formal institutions to handle. Even microfinance institutions require a group dynamic for savings and lending to the poor.

The technology for going digital and enhancing financial inclusion is sound and proven, the concept is appealing to both policy makers and donor communities, and service providers have found it to be a viable business case.

However, successful adoption and deployments are still few and far between. To date, cash remains the preferred medium of transactions in developing countries and even in many developed countries. So what is holding back application of digital systems and e-money for the delivery of affordable, accessible, and appropriate financial products, especially to people who are excluded from or underserved by formal financial markets?

Almost all of the transformational technologies are proven and available as off-the-shelf solutions that need little or no customization, at affordable prices. Hence, it is reasonable to expect that successful deployments—such as the M-Pesa mobile money platform in Kenya and social grants transferred through a biometric-enabled debit MasterCard in South Africa—could easily be replicated. However, such expectations have often met with disappointment when restrictions, insufficient underlying conditions, or other challenges constrained attempted deployments from achieving the kind of traction shown by those success stories.

Digital finance and payment services have transformed the lives of many people across the globe. *The Mobile Economy 2015*—a report by the London-based Groupe Speciale Mobile Association (GSMA)—counted 255 live mobile money services across 89 markets as of December 2014. The number of registered mobile accounts also grew to reach 299 million, with a majority of the accounts in Sub-Saharan Africa (figure II.1).

Nevertheless, the full potential of mobile money deployments remains untapped in some regions, such as Latin America and South Asia, partly because of regulatory barriers. Realizing the role that nonbank mobile money providers can play in fostering financial inclusion and economic growth, an increasing number of regulators are establishing new regulatory frameworks for mobile money. In 47 out of 89 markets where mobile money is available, regulation now allows both banks and nonbanks to provide mobile money services in a sustainable way, the GSMA report found.

The main purpose of Part II of this volume is to understand which factors enable successful deployment of digital money initiatives and why so few countries have succeeded in becoming fast movers with tremendous results. The study investigates the extent to which barriers are unique to each market, e-money deployment mechanisms, client segments and country context, and what lessons can be applied more generally.

Through analysis and examination of the case study countries and their respective e-money deployments, Part II of this volume identifies a set of critical enablers that have shaped the successes by unlocking value that allows for transformational outcomes. "Critical enablers" are game changers in enhancing the effectiveness and efficiency of e-money schemes in a transformative manner. They could be either general social enablers or program-specific enablers that help overcome major barriers to service proliferation. The lessons learned can benefit countries that plan to launch or are already in the process of launching similar initiatives, by identifying early implementation priorities.

At the macro level, such enablers include progressive policy leadership and an enabling, nonprohibitive regulatory environment. At the meso and micro levels,

Figure II.1 Number of Active Mobile Money Services Worldwide, by Region, 2001–14

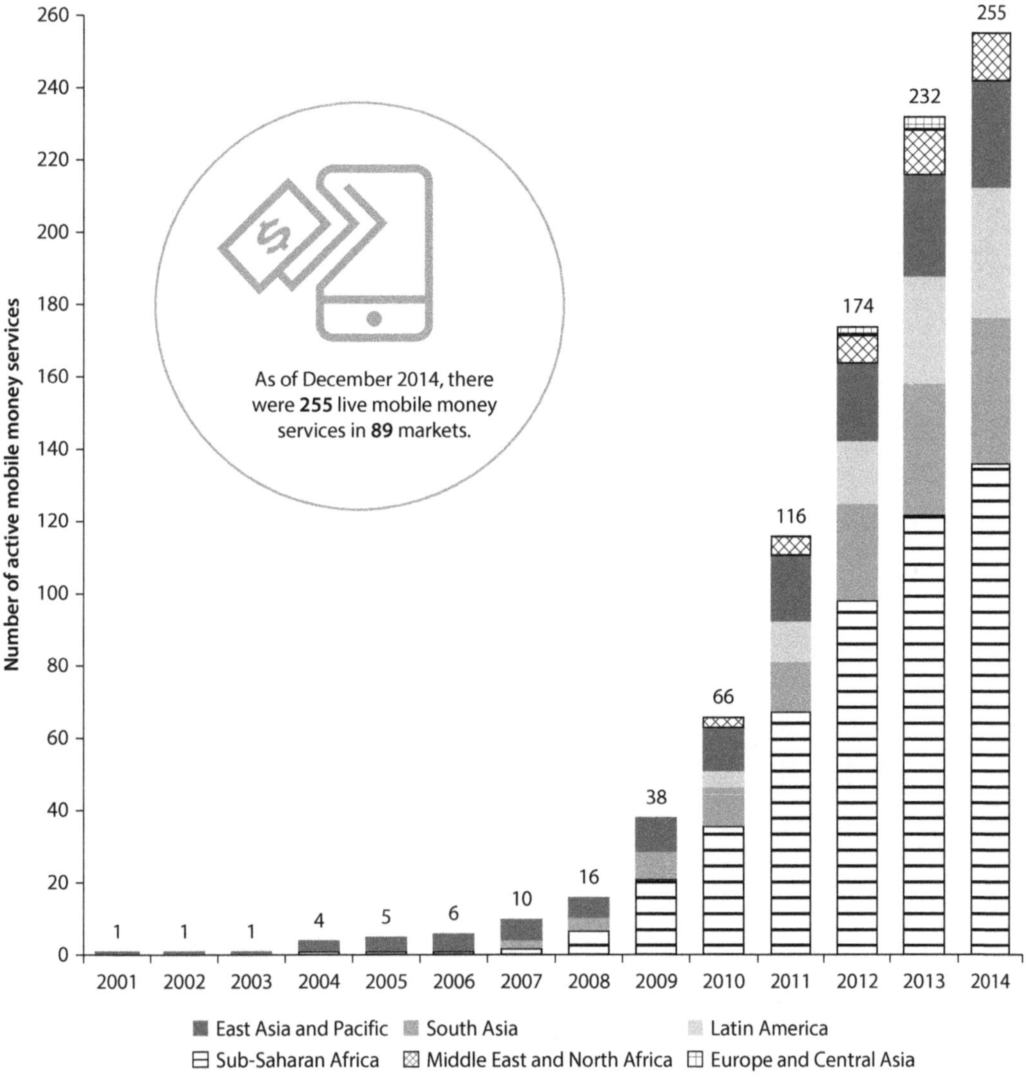

Source: Groupe Speciale Mobile Association (GSMA), *The Mobile Economy 2015* (London: GSMA, 2015). © GSMA. Reproduced, with permission, from GSMA; further permission required for reuse.

key factors include innovative use of infrastructure; an ecosystem characterized by interoperability; effective agent network management; mobile money add-on applications; and biometric-enabled, card-based grant payment disbursement systems. A basic requirement on the customer-level supply side is the ability to uniquely identify clients.

The case studies in this volume are used to illustrate how these factors have enabled some countries to rise above the rest in terms of ubiquity, speed, lower costs, simplicity, efficiency, and safety of their respective e-money deployments. In some cases, however, failure to address these key elements has hindered them from reaching the full potential of e-money deployments.

CHAPTER 4

Policy Leadership and Enabling Regulatory Environments

Introduction

According to World Bank 2014 Global Findex data, 2 billion largely poor people worldwide lack access to a financial transaction account, of which 30 percent live in three South Asian countries—Bangladesh, India, and Pakistan.[1] As such, governments and policy makers are prioritizing strategies and policies to make poverty reduction and economic development a reality, especially for the rural poor. Policy makers have to broaden their scope and commitment to make meaningful investments in responding positively to market innovations that open up new ways of delivering financial services. In doing so, while enabling inclusion strategies, policy makers will have to also ensure compatibility with financial stability and consumer protection.

Achieving financial inclusion requires greater emphasis and understanding of the poor and rural markets to establish a suitable enabling environment. Peake (2012) sums this up as follows:

> The rural customer segment has distinct characteristics compared with its urban counterpart. In most countries, agriculture and related activities represent a significant percentage of rural incomes, which typically result in seasonal flows and ebbs of income. Rural actors ... face greater constraints in terms of distance, travel times and infrastructure development ... so trust plays a huge role in engaging with them. Rural areas are also known to have lower literacy levels, lower mobile handset penetration rates and poorer network coverage. Finally, rural consumers are typically slower to adopt new brands and products but are also slower to give them up.

The factors cited by Peake apply broadly to people at the base of the pyramid, regardless of their geographical location. This understanding drives the policy leadership and regulatory efforts in certain developing countries to design and create policy and regulatory frameworks for innovative strategies to include the poor—or at times to allow for market developments in advance of regulatory or policy reforms.

To effectively harness the transformative potential of innovative digital financial products, mechanisms, and transactional networks, numerous legal and regulatory issues have to be addressed early on, including acceptable identification and validation of customer information; consumer protection in terms of information, funds, and dispute settlement processes; and how to apply legal and regulatory standards and guidelines. Sensible policy makers and regulators do not discourage these interventions by boxing themselves into preconceived regulatory mind-sets that overemphasize risk and returns or conform to conventional existing frameworks. Regulators in the more successful cases are open-minded and allow markets to come up with solutions that are workable, by allowing for pilot efforts that can be scaled up and then ensuring that broader regulatory parameters are appropriately adapted.

Regulatory Balance in Financial Innovation

Financial sectors are essentially a means to an end, facilitating financial intermediation to serve the real economy. Regulators of financial systems are entrusted with maintaining financial system stability, protecting consumer rights, and ensuring orderly market conduct. Although innovative financial systems and products have the potential to more efficiently allocate resources and support growth and development, certain innovations can bring in negative externalities and disrupt markets. As in the case of the global financial crisis, there is a risk of regulators scrambling to ring-fence the systems with draconian regulations.

An ideal regulatory balance would involve making complete information available for regulators to decide well in advance on the prudential norms needed to achieve safety and soundness, while market participants are allowed to be as innovative as they want to be. In reality, innovations may appear suddenly, and regulators may have to make a judgment call based on limited information.

With respect to policy leadership and an enabling regulatory environment, the analysis of the case study countries will answer key questions, as follows, involving prime examples of leadership:

- Why the Central Bank of Kenya (CBK) allowed M-Pesa to launch in a regulatory vacuum
- Why India's Jan Dhan Yojana program has the potential to be the flagship financial inclusion plan of the decade
- How Thailand's joint policy and regulatory vision contributes to enhanced financial inclusion in the country
- How Sri Lanka ensures that regulations keep pace with technological advancements that drive market developments
- How proactive policy leadership could enable two early movers—the Philippines and Maldives—to reach their full potential

Kenya: Leadership Lesson from the Central Bank of Kenya

The best example of an open-minded regulatory approach that led to transformative changes in poor people's lives is Kenya's M-Pesa mobile money platform. While much of the literature tries to explain why M-Pesa is hugely successful in Kenya and less so elsewhere—a one-of-a-kind success story that cannot readily be replicated—it is important to understand the key concepts that made M-Pesa a success.

Launch and Early Years of M-Pesa

The regulator's initial decision to allow the scheme to proceed on an experimental basis, without formal approval, was critical. Before the launch of M-Pesa in 2007, the CBK was aware of the woeful status of formal financial outreach in Kenya, which, at 19 percent, was highlighted by the FinAccess National Survey in 2006 (figure 4.1).

Soon after, when approached by the mobile network operator Safaricom, CBK realized the potential of leveraging mobile-based payments to enhance financial inclusion. CBK diligently reviewed all aspects of allowing mobile money in Kenya, while being mindful of the risks that this yet-unknown model might bring. Based on the details of M-Pesa's risk assessment and mitigation structure as well as opinions on the legality of allowing M-Pesa to function, CBK satisfied itself that

- M-Pesa could legally operate mobile-money business, and in doing so, it would not be conducting banking business;

Figure 4.1 Financial Access Strand in Kenya, 2006

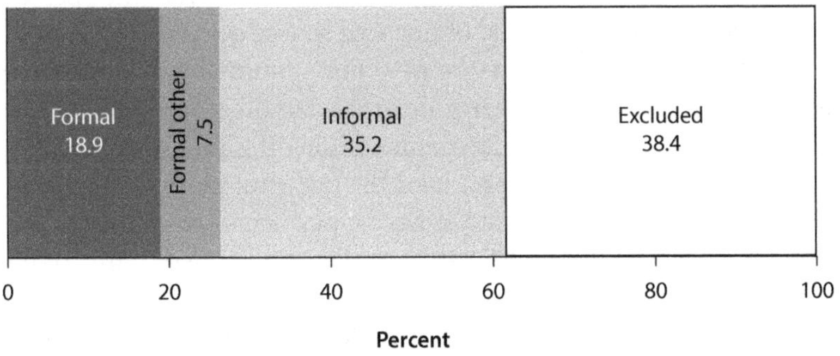

Source: FSD Kenya 2007.
Note: The figure reflects FinAccess 2006 Survey respondents who were ages 18 and older. The "Access Strand" is a graphical method (routinely used in FinScope Surveys in Africa) of placing each survey respondent along a continuum of access, depending on use of services by category. "Formal" refers to the share of the population using a bank, banklike institution, Postbank, or an insurance product. "Formal other" refers to the share using semiformal services from nonbank financial institutions (such as microfinance institutions) not bank services. "Informal" refers to those using only informal financial service providers such as rotating savings and credit associations (ROSCAs) and groups or individuals other than family and friends. "Excluded" refers to those using no institutionalized financial services.

- The risk management structure was adequate for consumer protection; and
- The pooled trust account in a reputable commercial bank, where money collected by Safaricom's agents is deposited, ensured that there would be no intermediation and also satisfactorily addressed the consumer credit risks.

CBK realized that, under the existing laws such as the Banking Act, they could not formally license or directly regulate the nonbanking institution on money transfer. However, it had full supervisory oversight of the pooled trust accounts that M-Pesa would hold with the commercial banks. Furthermore, the CBK Act (which established CBK in 1966) granted CBK broad powers to establish, regulate, and supervise payment and settlement systems. At the time CBK was considering the M-Pesa proposal, the National Payment Systems bill was still in draft mode. Instead of blocking the launch of M-Pesa until the National Payment Systems bill was passed, CBK issued a letter of no objection to Safaricom with conditions for consumer protection, prevention of money laundering, and regulatory reporting requirements.

M-Pesa was launched in March 2007 (see box 4.1). To its credit, CBK asked Safaricom to undertake two comprehensive technical assessments by Consult Hyperion[2] to evaluate the operational risk and efficiency of the M-Pesa platform; cleared the legal issues; and, once M-Pesa was launched, got a demand-side survey done through the Financial Sector Deepening (FSD) Kenya Trust in 2008. Although Kenya was not yet compliant with the Financial Action Task Force's (FATF) Anti-Money Laundering and Combatting Funding of Terrorism (AML/CFT) guidelines, Vodafone (the majority shareholder of Safaricom, which developed M-Pesa) ensured that the platform complied with its own international standards as well as with Kenyan legislation.

CBK came under tremendous pressure—from a disgruntled banking community, the political leaders, and the media—for allowing a nonbank to enter the payment space. The commercial banks were hostile to what they perceived as Safaricom's entrance into their field of financial service delivery. This hostility was exacerbated by the fact that the banks were not permitted to commence agency banking until 2010. However, having done its due diligence, CBK was able to convince the political leaders of the merits of using this unconventional route to enhance financial inclusion, and also used the opportunity to highlight the gaps in the regulatory system due to not yet having payments laws in place.

Thereafter, CBK and the Ministry of Finance issued joint statements informing the public of the due diligence process conducted by CBK and also gave assurance on the operational safety of M-Pesa. CBK relied on moral suasion and mutual cooperation with Safaricom. The CBK's National Payment System Department deliberately focused its oversight of M-Pesa—and of all subsequent mobile money services—on the integrity of the information technology (IT) platform and the service delivery system, and it continually monitors transaction flows and operations. Soon after M-Pesa's 2007 launch, CBK also conducted an audit of M-Pesa to confirm that M-Pesa is not a savings instrument.

Box 4.1 M-Pesa: A Backstory and an Alternative Perspective

Mobile network operator Safaricom was formed in 1997 as a fully owned subsidiary of Telkom Kenya—Kenya's sole provider of landline phone services. In May 2000, Vodafone Group PLC of the United Kingdom acquired a 40 percent stake and management responsibility for the company. In 2002, Safaricom was converted to a public company while the government held 60 percent of the shares, 25 percent of which would be auctioned off in 2008 on the Nairobi Securities Exchange.

The M-Pesa story begins with the U.K. Department for International Development's (DFID) formation of the FSD Trust in 2000. FSD awarded funds to Vodafone,

> which partnered with its affiliate Safaricom in Kenya to conduct workshops with financial institutions to identify the barriers to increasing access to financial services. A partnership was formed between a microfinance institution (MFI), a bank, and Safaricom to develop a pilot to enable microfinance loans to be paid with the help of mobile phones. (Cook 2015)

Funding from DFID through a competitively awarded challenge grant allowed a pilot to be launched taking a "test and learn" approach. In 2007, there were no regulations applicable to a mobile money service. The Central Bank of Kenya (CBK) took the brave step of allowing the innovation to move ahead, taking a risk-based approach. M-Pesa was launched in March 2007 focusing on its core money transfer function, marketed as "Send Money Home."

It is suggested that the government's bold regulatory decision to allow the launch of M-Pesa without the tight regulatory restrictions found in many other countries was driven by the following key factors:

First, the business model had been thoroughly piloted and tested as a project under the regulatory radar with flexible funding from FSD Trust.

Second, the government was the majority shareholder in the company that stood to benefit and had a dominant market position, meaning that

- Risk to the company from failure of the program was relatively small;
- Risk to the financial system likewise was relatively small, since the government was a backer; and
- Competitors were not in a strong position to lobby against the dominant, government-owned company—and they probably had no idea that M-Pesa would be so successful.

That CBK was willing to take a risk was also quite likely a legacy of the success of K-Rep Bank as a microfinance commercial bank, which had helped change attitudes in CBK as it supported legislation enabling MFIs to be licensed separately from banks.

Sources: Cook 2015; Safaricom information from Wikipedia, https://en.wikipedia.org/wiki/Safaricom.

All these good practices have enabled CBK to develop the necessary robust regulatory framework subsequent to commencement of M-Pesa's actual operation.

M-Pesa's Impact on Financial Inclusion

CBK's willingness to step outside of its comfort zone in the interest of a worthy public cause has unlocked great opportunities for poor people to better manage their money. Today, the financial inclusion rate in Kenya has risen to a remarkable 75 percent,[3] and M-Pesa is not the only operator active in the market. However, there are many things Kenya can do even now to ensure that the system functions smoothly, more affordably, and with a more level playing field.

According to the 2013 FinAccess Survey (FSD Kenya 2013), financial access is moving away from reliance on the informal sector, whose net addition to financial inclusion over formal financial institutions fell from 33 percent in 2006 to less than 8 percent in 2013 (figure 4.2).

Furthermore, the FinAccess Survey shows that mobile service provider use has achieved astoundingly rapid growth in penetration, from 0 percent of the population in 2006 to around 62 percent in 2013 (figure 4.3).

Figure 4.2 Financial Access Trends in Kenya, 2006–13

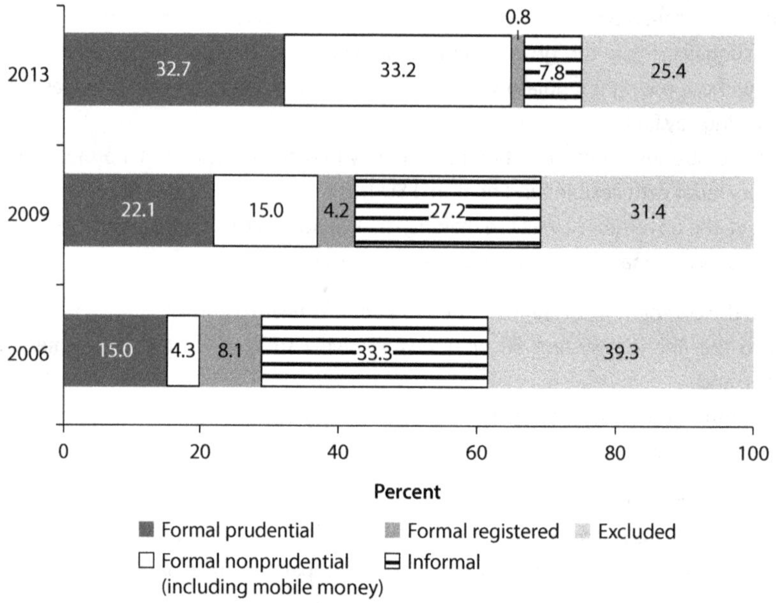

Source: FSD Kenya 2013.
Note: "Formal prudential" access refers to individuals whose highest reported financial services use is through providers that are prudentially regulated and supervised by independent statutory regulatory agencies. "Formal nonprudential (including mobile money)" access refers to those whose highest reported financial services use is through providers subject to nonprudential oversight by regulatory agencies or government departments or ministries with focused legislation. "Formal registered" access refers to individuals whose highest level of financial services use is through providers registered under a law or government direct interventions. "Informal" access refers to individuals whose highest reported financial services use is through unregulated forms of structured provision (FSD Kenya 2013).

Figure 4.3 Use of Financial Services in Kenya, by Type, 2006–13

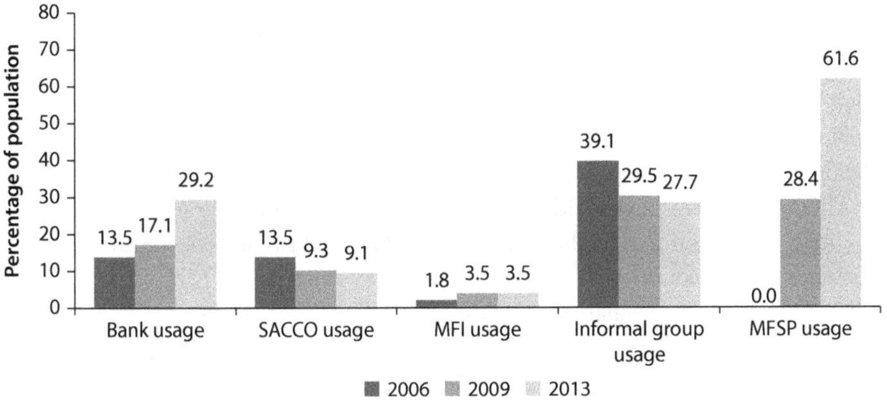

Source: FSD Kenya 2013.
Note: MFI = microfinance institution; MFSP = mobile-phone financial service provider; SACCO = savings and credit cooperative.

The Regulator's Journey since M-Pesa's Launch

The National Payment System Act came into effect in 2011. The focus of the Act's regulations was, however, on the large-value, high-systemic-risk payments through the CBK's Kenya Electronic Payment and Settlement System. CBK pretty much left the retail payments system to self-regulate.[4] Under the 2011 Act, CBK recognized and established a payment system management body with delegated powers to manage risks and governance issues and to support development of the national payment system.

In August 2014, a formal legal framework for mobile money was finally issued as the "National Payment System Regulations, 2014"[5] more than seven years after CBK's no-objection letter that facilitated the launch of M-Pesa. To help put things in perspective, during the 2014/15 fiscal year, M-Pesa had over 15 million active users transacting K Sh 4.2 trillion (US$4.32 billion) through the mobile money platform, equivalent to 42 percent of Kenya's gross domestic product (GDP). There is no question that if CBK had decided to uphold the status quo in 2006, most average Kenyan people would still be without access to the mainstream financial system.

India: Jan Dhan Yojana Flagship Financial Inclusion Plan

> *If crores of Indians are outside the ambit of organized financial services because they do not have a bank account even after 68 years of independence, I call it financial untouchability. Gandhiji ended social untouchability, it is our mission to eradicate this kind of untouchability now to fight poverty.*
>
> —Narendra Modi, prime minister of India, at the launch of Jan Dhan Yojana, August 28, 2014

Financial Inclusion Efforts in India

Financial inclusion is not a new concept to India. With over 1.25 billion people, financial inclusion has a special meaning, as over one-third of the world's poor live in India. Both the government and the Reserve Bank of India (RBI) have been pursuing this goal over the past several decades through building the rural cooperative structure in the 1950s, the nationalization of the banks in the 1960s, and the expansion of bank branch networks in the 1970s and 1980s.

These initiatives have paid off in terms of a network of branches across the country: by March 2013, there were also 93,488 Primary Agricultural Credit Societies in the country, 31 State Cooperative Banks, and 372 District Central Cooperative Banks (Sivaiah and Naidu 2015). Scheduled commercial banks also reached out into rural areas, with 49,571 branches and 504,142 Banking Correspondents in villages by 2015 (Patel 2016).[6] The major institutional innovation in India for expanding financial system access and use for the poor and marginalized sections of the population is the program that links banks with self-help groups.[7]

Yet, the extent of financial exclusion has remained staggering. Out of the 1.7 million total habitations (distinct clusters of houses at the subvillage level) in the country, only about 30,000 had a commercial bank branch as of 2012 (Chakrabarty 2012). Only about 40 percent of the population across the country had bank accounts. The proportion of people having any kind of life insurance coverage was as low as 10 percent, and the proportion having non-life insurance was an abysmally low 0.6 percent. People having debit cards comprised only 13 percent and those having credit cards only a marginal 2 percent.

As of 2011, only 46 percent of India's small and marginal farmers were able to obtain credit from either institutional or noninstitutional sources (Patel 2016). Even though RBI has mandated agriculture credit to be around 18 percent of the adjusted net bank credit, commercial bank data show that it was around 10.5 percent in 2013. Out of this, only about 4.5 percent reached the rural farmers, while the rest went to urban and suburban agriculture projects (Rao 2015).

The rating agency CRISIL, a Standard & Poor's company, has a financial inclusion index called the Inclusix. The all-India Inclusix score is 40.1 (which means that about 40 percent of the country has access to formal banking services). There are wide variations, from 62.2 percent in the southern region to 28.6 percent in the eastern region (Wharton School 2014). The 50 bottom-scoring districts have just 2 percent of the country's bank branches (CRISIL 2013).

The high-powered Nachiket Mor Committee on Comprehensive Financial Services for Small Businesses and Low-Income Households, set up by the RBI, found that 60 percent of the rural and urban population did not have a functional bank account (RBI 2014). One of the committee's recommendations was to set up a vertically differentiated banking system.

Accordingly, guidelines were issued by RBI and licensing effected for two new types of finance institutions in unbanked and underbanked regions: (a) "payments banks," which focus on remittance and payment services and accept demand deposits (current and savings) but cannot provide credit facilities; and (b) "small finance banks," which can offer a wide range of deposit (current, savings, and time deposits) and credit products, primarily microcredit, and also provide payment and remittance services.

The importance of this yet-to-be-proven payments bank concept is that it enables institutions with wide-reaching agent networks such as mobile network operators (MNOs) to offer payment transactions in a scaled-up fashion. Their challenge would be to match the already low-cost alternative transmission mechanisms that thrive in India and to attract customers to deposit with them. At some point in time, regulators will have to rethink the prohibition on payment banks offering credit products. The basic premise is that there has to be a business use case for them to proliferate.

Pradhan Mantri Jan Dhan Yojana: A Big Bang Approach to Financial Inclusion

Two weeks after his Independence Day address to the nation on August 15, 2014, Prime Minister Narendra Modi formally launched a mammoth financial inclusion program: "Pradhan Mantri Jan Dhan Yojana" (PMJDY), also known as Jan Dhan Yojana (JDY), which translates as "Prime Minister's People's Wealth Program." Touted as the biggest financial inclusion initiative in the world, the scheme aimed to take banking facilities to 75 million households within a period of five months, that is, by January 26, 2015.

By May 2015, banks had reportedly far exceeded the target set by the prime minister and opened approximately 159 million accounts, according to a survey of 210 million households in the country (table 4.1). In the first week of the scheme's launch—August 23–29, 2014—18.1 million accounts were opened, and the event entered the *Guinness Book of World Records*. Of the accounts opened, 60 percent are in rural areas and 40 percent in urban areas. The share of female account holders is about 51 percent.

The success of account openings was partly due to people getting a package deal when opening an account. Every account holder received a RuPay debit card, launched by the RBI-promoted National Payments Corporation of India (NPCI), with an inbuilt accident insurance coverage of Rs 100,000 (approximately US$1,650)

Table 4.1 Jan Dhan Yojana Account Status, by Bank Type, May 2015

	No. of accounts (millions)			*No. of RuPay debit cards*	*Balance in accounts*	
Bank type	*Rural*	*Urban*	*Total*	*(millions)*	*(Rs, billions)*	*Share of accounts with zero balance (%)*
Public sector banks	67.48	56.44	123.92	115.92	134.80	53.70
Regional rural banks	23.98	4.20	28.18	20.58	30.48	54.09
Private banks	3.85	2.68	6.53	5.88	9.92	49.65
Total	95.31	63.31	158.62	142.38	175.20	53.60

Source: PMJDY 2015.
Note: "RuPay" refers to a domestic debit card scheme launched in 2012 (initially called IndiaPay) by the National Payments Corporation of India as an alternative to foreign gateways with high transaction costs such as MasterCard and Visa and to consolidate various payment systems in India. Rs = Indian rupees.

and a RuPay Kisan credit card; and an overdraft facility of Rs 5,000 activated after six months of operation of the account. An incentive of life insurance coverage of Rs 30,000 was offered for those opening accounts before January 26, 2015 (celebrated as Republic Day in India) (*Economic Times* 2014).

Phase I of the scheme targeted three goals in the first year of implementation: universal access to banking facilities, provision of financial literacy, and provision of basic banking services. Phase II (August 15, 2015, to August 15, 2018) included the creation and development of a Credit Guarantee Fund to cover defaults in overdraft accounts and to provide microinsurance[8] and unorganized sector pension schemes. Coverage of households in hilly, tribal, and difficult areas and including every adult or student in all households is also planned.

The second phase is important because recent (2013) National Sample Survey Office (NSSO) data show that there are close to 57.7 million small-scale business units, mostly sole proprietorships, that undertake trading, manufacturing, retail, and other small-scale activities (Debu C 2015). Comparatively, organized sector and larger companies employ 12.5 million individuals. Today, the micro- and small-business segment is unregulated and without financial support from the formal banking system.

The promised Credit Guarantee Fund (Phase II activity) was launched in April 2015 as the Micro Units Development and Refinance Agency Ltd. (MUDRA) Bank, with capital of Rs 200 billion (US$3.13 billion) and a credit guarantee fund of Rs 30 billion to address the issue of lack of access to funds faced by small entrepreneurs. Although the government had considered evolving the MUDRA Bank to be the regulatory body for MFIs, this plan did not proceed. Nevertheless, MUDRA provides its services to small entrepreneurs outside the service area of regular banks, by using "last mile" agents. About 5.77 crore (57.7 million) small businesses have been identified as target clients using the NSSO survey of 2013. Only 4 percent of these businesses get finance from regular banks.

Challenges to Jan Dhan Yojana

One of the major concerns that haunts the Ministry of Finance is that about 54 percent of the accounts opened have zero balances (table 4.1). Criticisms also include that JDY was a public-bank-driven exercise. However, both of these trends seem to be waning (figure 4.4).

The banking sector is realizing that the JDY scheme can bring in huge opportunities in the years to come. Banks will be able to offer microfinance and related products to people across the country. The question is whether the scheme is likely to make rural operations more profitable for banks. As stated earlier, rural accounts make up 60 percent, and regional rural banks 18 percent, of the total JDY accounts opened to date (PMJDY 2015). The government has successfully addressed the challenge of getting accounts opened.

Through this initiative, savings from these typically low-income households are now brought within mainstream banking. However, getting customers to use these accounts is the second challenge. Because some of the offers become validated only once the account is operational, this is a matter for concern.

Figure 4.4 Zero-Balance Trends in Jan Dhan Yojana Accounts, India, 2014–15

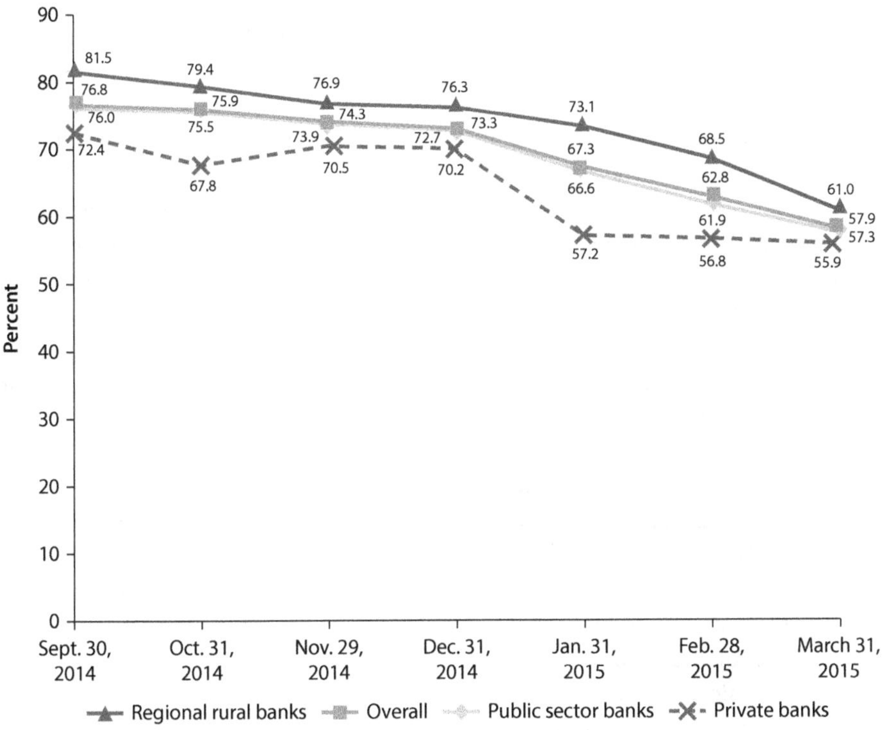

Source: PMJDY 2015.

The accident insurance will come into effect only if at least one transaction takes place within a 45-day period. The overdraft facility will depend on the balance in the account. Given the poverty and illiteracy among the target groups, a comprehensive awareness campaign is a must. The trend in zero-balance accounts is downward-sloping, which is encouraging (figure 4.4). Hence, it looks as if the government's awareness campaign is working.

Direct Benefit Transfers: Key Enabler for Jan Dhan Yojana

The potential game changer for financial inclusion could be the direct benefit transfers (DBTs) channeled through JDY bank accounts. The government is hastening this process by linking all accounts to India's Aadhaar unique identification (ID) program. Once this is completed, the flow of money from government schemes to the beneficiaries will become seamless.[9] The bank will also enjoy the benefit of the float if all government payments to beneficiaries are transferred to these accounts. In addition, there is a proposal that the government pay the banks a 2 percent commission on the total DBT payments handled. However, the banks are negotiating for 3 percent plus a tax adjustment. Whatever the negotiated amount finally agreed, it will have to make business sense for the banks if efficiency is to prevail.

The saving on leakages that occur in various subsidy schemes will more than compensate for the cost of this massive exercise of financial inclusion. Leakages are estimated to be between Rs 500 billion and Rs 750 billion (around 5 percent of GDP).

Realizing that incentives are needed to keep these accounts from remaining or turning dormant, the government is inclined to channel all benefits, including state and local government benefits, through these accounts. Though DBTs and the mandate for continued eligibility for accident insurance coverage are incentives to operate the bank account, there is nevertheless the risk that many accounts not covered under DBTs may fall into dormancy. Even for accounts covered by DBTs, operations may be confined to receiving the DBT remittances and withdrawal. However, notwithstanding the realities of how poor people operate basic no-frills accounts, the potential for a program to bring millions of unserved people into the formal financial system is a remarkable feat.

Digitizing the Last Mile Connectivity

The biggest challenge for the JDY program is to deliver financial services using a sustainable and scalable model. Even though the government and RBI have taken measures to expand the access channels, a scaled-up digital payment system is not yet widely available in India. Hence, the full potential of digital financial services in India has not yet been realized.

However, since 2006 in Andhra Pradesh, social welfare payments have been disbursed electronically, through around 26,000 female self-help group members, to 16 million accounts. In fiscal year 2012/13 alone, the government of Andhra Pradesh channeled US$1.2 billion worth of social welfare payments through electronic channels, and this is expected to increase. Andhra Pradesh was the first state in India to implement an electronic payment channel at scale (CGAP 2013). With the DBT scheme rolling out countrywide, there are lessons to be learned from the initial efforts in Andhra Pradesh.

Given the inadequacies in banking infrastructure as access points to the general public, digital payments and the payment banks can play a significant role in making financial access a reality. In addition, using India Post—with 155,000 post offices, of which nearly 90 percent are in rural areas—is also considered an effective way to provide service touchpoints.

One significant shortcoming that is also a huge opportunity is that India lags significantly behind in mobile payments. The Consultative Group to Assist the Poor (CGAP) reported that control by the banks of not only credit and savings but also the payments market has restricted the use of mobile money to only 0.3 percent of adults in India (Kumar and Radcliffe 2015). With 900 million mobile connections in 2013 across India—expected to rise to 1.16 billion by 2017—mobile connectivity presents the perfect opportunity to increase the use of e-money (figure 4.5). According to the Telecom Regulatory Authority of India (TRAI), there were 350.37 million rural subscribers by 2013 (TRAI 2013). The MNOs already manage a distribution infrastructure and large agent network of 2–2.5 million touchpoints, and they

Figure 4.5 Number of 2G and 3G/4G Connections in India, 2008–17

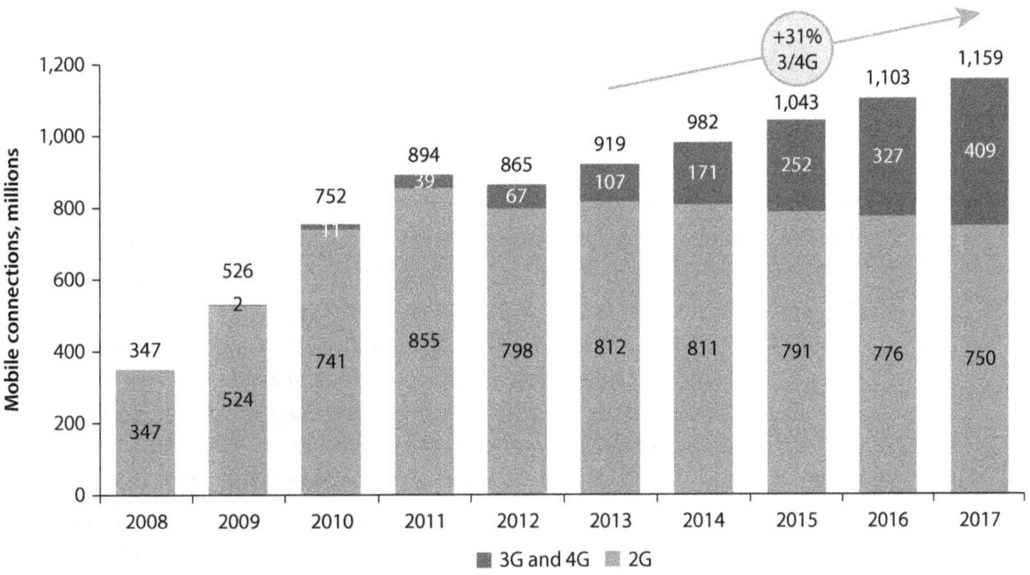

Source: MMAI and GSMA 2013.
Note: 2G, 3G, and 4G refer to the "generation" of the mobile-phone network; the higher the number, the faster the download speed as a measure of how quickly information can be transferred from the Internet to a phone. 2G mobile networks are handled by simple cell phones for basic e-mails and texts. These second-generation or 2G phones, which are also low-cost and suitable for poor people worldwide, are quite adequate for mobile money applications that operate by sending short message service (SMS) or texts.

already address the needs of low-income consumers by designing products such as Rs 10 (US$0.16) airtime reload vouchers (MMAI and GSMA 2013).

Realizing this potential, in 2014 RBI brought in transformative regulatory reforms that enable the use of this untapped network of agents to expand the transaction points across India so that people can readily access basic banking services and potentially cash out their subsidy payments without incurring additional cost in terms of time as well as money (box 4.2).

In addition to RBI regulations, TRAI issued guidelines that require mobile operators to provide banks with unstructured supplementary service data (USSD) channel access for mobile banking. This will ensure that banks can leverage mobile operators' communications channels when offering mobile banking (Kumar and Radcliffe 2015). To facilitate greater adoption of mobile banking, RBI recommends that all MNOs preload a standard single mobile-banking application in the form of a short message service (SMS), and the government has mandated this addition to all new phones.

The NPCI also plays its part in this important initiative by setting up a next-generation USSD-based mobile banking service: the National Unified USSD Platform (NUUP). Accessible through a common code (*99#) dialed from a mobile phone, NUUP allows every customer to access banking services with a single number across all banks—irrespective of the telecom service provider, mobile handset make, or the region. However, NUUP is only available on Global

Box 4.2 Reserve Bank of India Regulatory Reforms, 2014

- Payments bank guidelines are to allow companies with significant distribution expertise (including mobile operators, retail chains, and existing agent managers) to offer deposit accounts and payments as a stand-alone business.
- Reserve Bank of India (RBI) lifted its prohibition against banks establishing agents more than 30 kilometers from the nearest bank branch. The 30-kilometer rule has befuddled the financial inclusion sector for years by preventing smaller banks with limited branch networks from building national agent networks. By lifting this restriction, the RBI effectively leveled the playing field between large and small banks, at least when it comes to agent banking. And once payments banks come online, they will be able to establish agents without worrying about building brick-and-mortar branches to comply with the 30-kilometer rule.
- RBI removed the requirement that customers provide proof of current and permanent address for opening a bank account. This requirement was particularly tricky for migrant laborers who have trouble securing documents that prove their current address.
- Nonbanking finance companies can now act as business correspondents. This will allow India's microfinance institutions—many of which already serve millions of poor customers—to build agent networks on behalf of banks.
- RBI concluded its pilot to determine whether licensed prepaid issuers, such as Airtel Money and Vodafone M-Pesa, could allow their customers to cash out. RBI has removed that restriction, creating more options for nonbank providers to offer payment services.

Source: Kumar and Radcliffe 2015.

System for Mobiles (GSM) mobiles and not on Code Division Multiple Access (CDMA) mobiles.[10] A platform common to all banks—instead of each bank having to develop its own platform—is a strategic move by NPCI to help banks focus on customers while NPCI manages the technology behind the platform. NUUP is currently available in 11 languages. NPCI and the regulators successfully brought in all 12 MNOs.

NUUP is using the Immediate Payment Service (IMPS) platform, which is a real-time interoperable payment mechanism by NPCI. IMPS offers an instant, 24-hour, 7-days-a-week interbank electronic funds transfer service through mobile phones. IMPS transfers money instantly within banks across India through mobile, Internet, and automated teller machines (ATMs), which is not only safe but also economical both in financial and nonfinancial terms. There are no charges on NUUP transactions. However, IMPS fund transfer charges would be applicable.

Sri Lanka: Regulations Keeping Pace with Technological Advancements

The dawn of the Mobile Money era in Sri Lanka has been made possible by the progressive, and financial inclusion focused, regulatory ethos of the Central Bank of Sri Lanka. In this respect Sri Lanka's Payments and Settlement Legislation and Mobile Payment regulations stand among the most progressive in the world. Emboldened by the support and facilitation received from the Central Bank, Telecommunication Regulatory Commission and our custodian bank [Hatton National Bank], we are confident that this pioneering initiative to deliver a nationwide Mobile Money will deliver an unprecedented level of financial inclusion through the empowerment of millions of Sri Lankan citizens with electronic money transfer and payment facilities.

—Hans Wijayasuriya, group chief executive, Dialog Axiata PLC, at the launch of eZ Cash, June 2012

A good financial sector regulatory framework should be simple and effective, protect consumers, and reward innovation by being flexible enough to adapt to changes in the markets, while combatting money laundering and fraud. More often than not, frustrated innovators claim that regulators cannot keep pace with the rapid changes and technological developments in the industry, while skeptical regulators maintain that such innovations may do more harm than good. Often it is a balance between these two extreme viewpoints that would make a good enabling regulatory system that fosters responsible innovation. Risk management systems should keep pace with the complexity of new financial products to ensure that there are no pervasive failures in consumer protection; transparency and standards should be maintained to allow for fair market conduct and competition; and gaps and weaknesses in the supervision and regulation of these newcomers to the market may have to be addressed in innovative ways.

Mobile money systems thrive on markets where banking density is very low or people are geographically widely dispersed, thus making financial access a difficult task. Sri Lanka has no serious issues in either category. Nearly 70 percent of the adults in Sri Lanka have accounts in formal financial institutions,[11] and in their report "Deposit Assessment in Sri Lanka," Leonard et al. (2011) noted that outreach of financial services in Sri Lanka can be considered fairly extensive, with a reported 82.5 percent of households having access to financial institutions for their savings and credit needs.

Since the 26-year civil war ended in 2009, the government and the Central Bank of Sri Lanka (CBSL) have spearheaded financial inclusion activities focusing on the war-torn areas of the North and East emerging from the conflict. For example, banks were required to open two branches outside of the Western Province for every branch opened within the Western Province. By 2014, the Northern Province had the highest banking density (21.7 branches per 10,000 people), and the Eastern Province the third highest (16.8 per 10,000) (*Daily FT* 2015a). At the same time, the CBSL authorized testing of a mobile banking service, leading to the launch in 2012 of eZ Cash, which gained over a million customers within a year (di Castri 2013). Hence, in Sri Lanka, mobile money solutions offered by MNOs will always be one of many options available to access financial services, and will proliferate only if there is a simple and clear value proposition to the users.

CBSL decided to allow mobile money services through MNOs after making sure that necessary safeguards are in place. It is a trailblazer in enabling technological innovations by setting up benchmark regulatory practices that are both progressive and nonprohibitive. CBSL enacted necessary laws early on. It created a level playing field where both banks and nonbanks including MNOs can offer digital money; established enabling regulatory and oversight functions that are not overtly restrictive; ensured that customer rights and information are well protected; and worked with stakeholders across the board to understand market needs and innovations. These positive reforms did not happen overnight.

Legislative Background

Being a pioneer in the South Asia region to adopt technology-based financial services, Sri Lanka has broader legislation than its regional neighbors for payment systems and electronic transactions—and a wide range of payment instruments and payment services to support the country's banking and finance industry. CBSL has always been at the helm, providing leadership, guidance, and a legislative framework to enable such developments.

The most significant decade in terms of regulatory enactments relating to payment reforms and e-money in Sri Lanka has been the 2001–10 period. Under the central bank's modernization project that commenced in 2001, CBSL's objectives have been redefined by an amendment to the Monetary Law Act in 2002 to recognize financial system stability as a statutory objective in addition to the price and economic stability objective. Starting with this momentous amendment, a series of other important financial regulatory laws were enacted during this period.[12] Additionally, several amendments were introduced to the existing laws to facilitate new payment and settlement systems and processes (Jayamaha 2014).[13]

The rapid pace of the passage of these laws indicates the proactive leadership of CBSL in establishing the core legal infrastructure required to facilitate the developments of electronic transactions and mobile payments. As part of the payment systems modernization undertaken in 2001, to improve efficiency and enhance the level of integrity, the Payment and Settlements Systems Act (PSSA) was enacted as the comprehensive legislation governing payment, clearing, and settlement systems. The Act provided CBSL with oversight and regulatory powers over the national payment system as well as payment systems and money transfer service operators. PSSA was the key legislation that enabled nonbanks to offer mobile money services.

The Electronic Transactions Act of 2006 provided the underlying framework for electronic and mobile transactions, based on the model laws developed by the United Nations (UN) Commission on International Trade Law, which recognizes and enables enforceability of electronic transactions, and the UN Electronic Communications Convention. Three fundamental policy-related principles form the basis of the Act: (a) technology neutrality, (b) functional equivalence, and (c) party autonomy. Technology neutrality is ensured by not dictating the technology to be given legal preference. Hence, unlike statutes in some countries, the Electronic Transactions Act in Sri Lanka does not specify any technology that should be used, allowing businesses and consumers to determine technology options based on types of use.

Sri Lanka became the first country in South Asia (and one of the first three in the Asian region, along with China and Singapore) to sign the UN Electronic Communications Convention in July 2006 (Fernando 2013). In terms of the provisions of this Act, all transactions and business done in electronic form would be recognized and valid, with the exception of excluded items (under section 23 of the Act) such as last wills, powers of attorney, and transfers of immovable properties, to name a few. This Act also facilitated Sri Lanka's e-government services available through the Lanka Gate web portal.

In July 2015, Sri Lanka ratified the UN Electronic Communications Convention, another first for South Asia and the second country after Singapore to become a State Party to the Convention. Countries such as Australia, China, Thailand, and Vietnam are already preparing domestic legislation to ratify this Convention. Jayantha Fernando, legal adviser to Sri Lanka's Information and Communication Technology Agency (ICTA), said this of the action: "Sri Lanka's ratification of this Convention will ensure greater legal certainty for e-commerce and e-business providers who wish to use Sri Lankan law as the applicable law and ensure international validity for other international legal instruments as well as cross-border fund transfers, enhancing the ability of Sri Lanka to fast-track its move towards paperless trade facilitation" (*Daily FT* 2015b).

Sri Lanka has also adopted the recommendations of FATF and enacted several laws enabling CBSL to issue AML/CFT guidelines and regulations to all financial sector players and payment systems providers.[14] Furthermore, realizing the need for greater information security, the Payments Devices Frauds Act (No. 30 of 2006) and the Computer Crimes Act (No. 24 of 2007) provided a unique investigation and enforcement regime. More recently, Sri Lanka became the first country in South Asia to be invited to join the Budapest Convention on Cybercrime.[15] Although the much-needed Secured Transactions Act (No. 49 of 2009) was enacted with the intention of establishing an online registry for movable assets, amendments are needed to include new provisions to enable movable and mobile collateral to be accepted as sureties.

Institutional Infrastructure

The ICTA of Sri Lanka, the government's apex information and communication technology (ICT) institution, ably supported CBSL in establishing key legislation and a comprehensive information security framework for the financial sector to ensure greater consumer confidence as the financial sector transitions toward a cash-lite electronic mode. ICTA addressed the legal validity and enforceability of electronic transactions—the main concerns for the financial community in transitioning to electronic mode—by spearheading the enactment of the Electronic Transactions Act in 2006.

CBSL has empowered LankaClear to be the national payment infrastructure provider for retail payments.[16] Driving Sri Lanka toward an efficient, green, and paperless nation, LankaClear provides the necessary technological platforms and gateways that are cutting-edge and cost-efficient and allow for seamless connectivity. Some of these are groundbreaking and the first in South Asia. These solutions are both efficient and secure.

Among them, the LankaSIGN certification service provider (CSP) implemented by LankaClear in 2009 provides the much-needed security for the electronic payment systems and is currently the only commercially operated CSP in the country. Development of the Common Card & Payment Switch (CCAPS) by LankaClear under the guidance of CBSL is an important ongoing initiative in the retail payment arena.

LankaClear also hosts the Bank Computer Security Incident Response Team (Bank CSIRT). Bank CSIRT, a joint initiative of CBSL and the Sri Lanka Computer Emergency Response Team (SLCERT), is a specialized service unit that is responsible for receiving, reviewing, processing, and responding to computer security alerts and incidents affecting the banks and other licensed financial institutions in the country. Commercial banks provide guidance and funding for Bank CSIRT. Through LankaClear, Bank CSIRT will have the prime responsibility to coordinate security efforts within the banking and financial sector.

In addition, a special unit called the Cyber Crime Investigations Unit is established in the Criminal Investigations Department of the Sri Lanka Police Service. It is reported that CBSL and SLCERT are jointly taking the initiative to prepare legislation and regulations necessary to deal with cybercrimes.

As a result of this legislative framework, Sri Lanka has an excellent payment systems infrastructure and possibly the best financial regulatory framework in the region to govern e-money for e-commerce and e-government programs. In the World Bank's 2008 Global Payment Systems Survey, the legal and regulatory framework of Sri Lanka's payments system was rated at the "High Level of Development" (Cirasino and Garcia 2008; World Bank 2008). The highest ratings were given for system design and key policy decisions that affect the safety, soundness, and efficiency of the system.

Notwithstanding the excellent institutional infrastructure that enables innovative solutions, the most important institutional arrangement from the mobile money perspective is the coordinated regulatory collaboration between CBSL and the Telecommunications Regulatory Commission of Sri Lanka (TRCSL). The expansion of digital technologies has altered the ways in which financial services are delivered and accessed. Thus, regulators are presented with a number of challenges, specifically regarding consumer protection and due diligence.

The uniqueness of e-money is that it overlaps different regulatory domains, thus risking a mismatch of regulations and potential coordination failures. By coordinating proactively and by converging the service delivery and customer protection paradigms of the telecommunications and banking sectors, the two regulatory authorities in Sri Lanka were able to provide a comprehensive yet flexible mobile money regulatory and oversight framework where CBSL would take the lead. TRCSL is agreeable and pledges support to cooperative efforts to ensure smooth implementation of regulatory initiatives.

Leveling the Playing Field

CBSL realized early on that an effective, inclusive digital financial system requires a level playing field so that private sector players such as MNOs can offer competitive products and services efficiently and at a lower cost, using their ready-made network of reload agents as touchpoints. While not creating unnecessary hurdles for the MNOs, and ensuring that safeguards are met, CBSL decided on an iterative approach.

Unlike in most other countries, the already-enacted PSSA in 2005 gave CBSL the necessary powers to authorize MNOs to offer mobile money. In 2007, CBSL authorized the National Development Bank (NDB), a smaller commercial bank, to issue a mobile money service jointly with Dialog Axiata, the leading MNO in the country,[17] which wished to pilot a bank account–based mobile money system. The regulatory framework required the customers to have a bank account to be able to sign up for this mobile money service called "eZ Pay" (subsequently upgraded to "eZ Cash"). Although two more banks and a registered finance company also joined, the system did not scale up easily, and it ended up having only around 15,000 customers.

In this case, even though Dialog Axiata operated under NDB to offer the mobile money service, NDB required the same level of identity verification from mobile money clients as from bank customers. The success of mobile money and scaling-up depends on the ability to speed up the wallet account opening process and to minimize steps in transacting. Realizing that customers were not interested in signing a plethora of documents to verify their identity in signing up for bank accounts, Dialog Axiata asked CBSL to consider an MNO-led mobile banking solution without requiring a bank account to be opened by customers. CBSL reexamined the guidelines and reviewed the submissions by Dialog Axiata to operate an MNO-led mobile money solution using the customer information already available in their mobile-phone contracts. Even though the yet-unfolding global financial crisis and the collapse of a local credit card company were, at the time, putting pressure on the regulator, CBSL was aware that limiting innovation and competition would result in suboptimal noncompetitive solutions that would affect financial inclusion as well as financial system stability in the long run.

At the same time, consumer protection and information safeguards were primary considerations. Addressing all these concerns, in 2011 CBSL issued two distinct mobile money guidelines, for the bank-led model and the MNO-led model under PSSA.[18] Accordingly, mobile money or payment cards service provider licenses could be granted to

- A licensed commercial bank;
- A licensed specialized bank;
- A finance company;

- An operator that provides cellular mobile telephone services under the authority of a license issued under the Sri Lanka Telecommunications Act (No. 25 of 1991) as amended; or
- A company registered under the Companies Act (No. 7 of 2007) having an unimpaired capital of at least Rs 150 million or such other amount determined by the Central Bank, other than a company limited by guarantee, an offshore company, or an overseas company, within the meaning of the Companies Act.

By issuing these comprehensive and broad guidelines under PSSA, CBSL opened up the horizon for a broad and growing range of providers with different value propositions to participate in the payment services industry.

Proportional, Risk-Based Know-Your-Customer Requirements

Unlike banks and traditional payment methods, mobile payments pose regulatory challenges in addressing AML/CFT risks. Such risks stem from the lack of face time with customers, leading to difficulties in identification and verifying income sources. Dealing with MNOs and transacting across a vast number of agents who traditionally are not from the financial sector also pose challenges for financial institutions and regulators.

On the other hand, overly strict requirements regarding the identification and verification of customers risk exacerbating financial exclusion. FATF recommendations encourage the use of (a) differentiated customer due diligence (CDD) measures according to the profile of the (future) customer for customer identification; and (b) "progressive" or "tiered" CDD as a risk-based "know-your-customer" (KYC) method. This would mean that people lacking a government-issued ID or formal proof of address can be allowed basic access with limited functionalities using alternative identification methods. Access to a wider range of services and higher transaction limits may be allowed when additional CDD requirements are met. Hence, transaction and payment limits or wallet sizes vary based on the KYC or CDD requirements met.

Sri Lankan citizens who are 16 years and older and reside in Sri Lanka are required to apply for the national identity card (NIC) issued by the Registration of Persons Department. The NIC is regarded as the key ID document for identification and authentication of persons. Also in Sri Lanka, subscriber identification module (SIM) registration is mandatory, hence MNOs already have copies of national ID cards and photographs as well as digitized versions in their databases.

Considering these facts, CBSL applied the proportional risk principle that allows low-value mobile wallets to be automatically enabled via the customer dialing in using the customer information already available to the MNOs in the mobile contract. Higher-limit mobile wallets would require customers to visit an MNO kiosk and provide additional CDD-related paperwork. This means that almost all mobile subscribers (22.12 million subscribers, or 107 percent of the population in Sri Lanka) are able to operate a basic e-wallet if they so wish, just by dialing in. A low initial registration burden for basic wallet size reduces customer resistance to registration and hence enables faster growth in the use of mobile money.

Regulatory Oversight

Due to the very nature of mobile transactions, mobile money initiatives are deemed to be low-value, low-risk systems. Hence, oversight and monitoring of service providers is not expected to be as extensive as supervision conducted for banks and other financial institutions that are also subject to prudential norms such as the Basel Core Principles (BIS 2012). However, the magnitude of mobile money proliferation in Kenya suggests that such initiatives have the potential to become systemically important payment systems.

Because of the innovative nature of these initiatives, often the laws and regulations have gaps and gray areas that create oversight challenges and may lead to negative reactions by the regulatory authorities. For example, in 2014, China's central bank suspended mobile payments initiated through Quick Response (QR) codes amid security concerns regarding the identification process involved with those transactions. The bank's decision immediately affected China's two largest third-party mobile payments providers: Alibaba Group Holding Ltd. (which operates Alipay) and Tencent Holdings Ltd.[19] In Kenya, M-Pesa functioned for more than seven years without a regulatory framework before CBK was able to pass the National Payment System Regulations in 2014.

In Sri Lanka, having the PSSA enacted in 2005 afforded greater clarity and transparency to the oversight role of CBSL, and the Act legally empowered the CBSL to regulate, supervise, and monitor the payment systems and service providers. Thus, understanding the need for legal certainty,[20] CBSL issued regulations under PSSA to license the mobile payment service providers.[21] Thereafter, mindful of the innovative nature of mobile money services and lessons learned from previous mobile money deployments that failed to gain traction, CBSL issued the Mobile Payments Guidelines (No. 2 of 2011) for Custodian Account Based Mobile Payment Services, which provides the monitoring framework for nonbank service providers to operate. Issued under the PSSA, these regulations and guidelines ensure that the regulation and oversight function supports payment system stability and also instills public confidence in mobile money services. The reporting requirements and regulatory and oversight provisions are clearly set out in the guidelines. Moreover, CBSL has also entrusted the custodian bank of the licensed service provider with the additional responsibilities of formulating KYC and CDD procedures for the service provider, monitoring and supervision of the service provider for compliance with regulations and guidelines, and auditing of all e-money accounts with the service provider (an MNO in this case).

Customer Protection

In line with good-practice prudential requirements, CBSL requires the MNOs to "ring-fence" the customers' funds in custodian accounts at licensed commercial banks.

According to the guidelines issued, it is mandatory that the e-money accounts be updated by the service providers on a real-time basis and that the cumulative sum collected from all e-money account holders be maintained in the custodian accounts at all times.

Additionally, a trust is established by execution of a trust instrument administered by Deutsche Bank AG over the funds held in the custodian account. Hence, in the event that either the service provider (the MNO) or the custodian bank becomes insolvent, customer funds in the pooled account will be protected.

Customer Education and Grievance and Redress Mechanisms

The regulators should ensure that service providers educate customers about using security features and the importance of protecting their personal information. Hence, there should be transparent, effective, and straightforward complaint mechanisms and recourse. To this end, CBSL has directed the service providers to develop an appropriate dispute resolution mechanism based on guidelines that will be issued by them to handle disputed payments, transactions, and loss of mobile phones. Service providers are required to establish a call center to respond to customer inquiries and complaints and to be responsible for addressing customer grievances in the event a customer files a complaint about a disputed transaction. Charge-back procedures for addressing such customer grievances are to be formulated by the service providers.

The early enactment of PSSA allowed CBSL to establish focused yet flexible regulatory mechanisms to regulate e-money. The cornerstones of this successful regulatory framework were a nondiscriminatory licensing regime; consumer protection and safeguarding of funds; reduction of operational risks and establishment of tiered KYC and CDD measures; and appropriate oversight mechanisms augmented by use of prudentially regulated custodian banks to monitor and assist with e-money operations of the nonbank service provider. Ably supported by an enabling regulatory framework, eZ Cash has the distinction of being the first mobile money system in the world to be end-to-end interoperable across multiple service providers, thus bringing forward a combined subscriber base of over 14 million Sri Lankans who could transact electronically via eZ Cash.

Thailand: A Government's Vision and Policy to Bring Cash to the Doorstep

According to the 2013 FinScope Survey, in Thailand 97 percent of the adult population have accessed formal financial services: 74 percent of the adult population have bank accounts, while an additional 23 percent have used formal nonbank financial services (FinScope 2016). With another 2 percent of the population using informal financial services, the financially excluded population in Thailand is only 1 percent (figure 4.6).[22] These ratios are similar in all

Figure 4.6 Financial Access Strand in Thailand, 2013

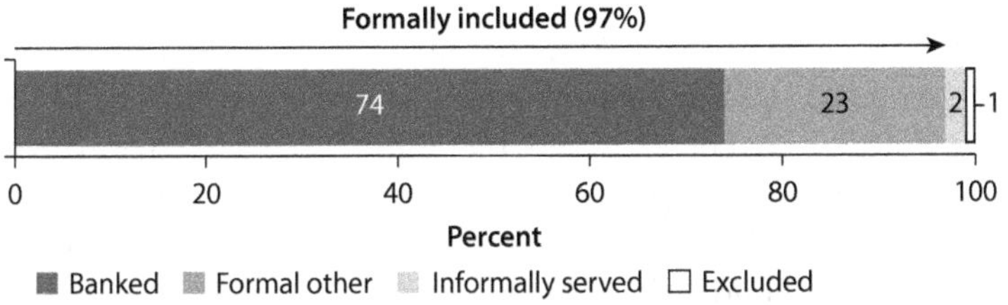

Source: FinScope 2016.
Note: The figure reflects FinScope 2013 Survey respondents who were ages 18 and older. The "Access Strand" is a graphical method (routinely used in FinScope Surveys) of placing each survey respondent along a continuum of access, depending on use of services by category. "Banked" refers to those who reported using traditional financial products supplied by commercial banks. "Formal other" refers to those who reported using financial products of formal financial institutions that are not commercial banks. "Informally served" refers to those who reported using "informal" products without recognized legal governance, such as savings with an employer or savings group. "Excluded" refers to those who reported using no financial products.

Figure 4.7 Financial Access Strand in Thailand, by Region, 2013

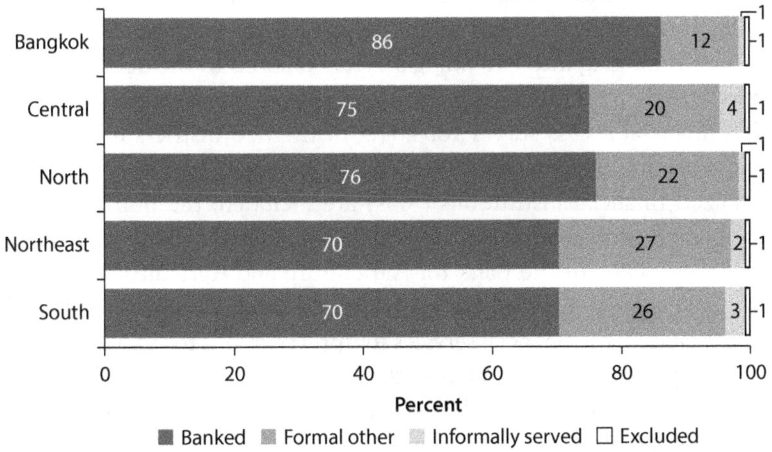

Source: FinScope 2016.
Note: The figure reflects FinScope 2013 Survey respondents who were ages 18 and older. The "Access Strand" is a graphical method (routinely used in FinScope Surveys) of placing each survey respondent along a continuum of access, depending on use of services by category. "Banked" refers to those who reported using traditional financial products supplied by commercial banks. "Formal other" refers to those who reported using financial products of formal financial institutions that are not commercial banks. "Informally served" refers to those who reported using "informal" products without recognized legal governance, such as savings with an employer or savings group. "Excluded" refers to those who reported using no financial products.

four regions of the country, with the Bangkok area showing higher access through banks at 86 percent (figure 4.7).

The high level of access can be attributed to the deliberate policies pursued by the Royal Thai government (RTG) over the years to extend financial services to the underserved and unserved population. These policies have been largely government-led and government-financed. The commercial sector has played a relatively minor role in providing access to credit and insurance, while taking the lead in savings and remittances (ADB 2013a).

One government strategy was to involve state-owned financial institutions, such as agricultural cooperatives and the Village and Urban Revolving Fund (Village Fund). The Village Fund is a semiformal institution established in 2001 by the RTG after the 1997 financial crisis to make credit available to the villages. The Thailand Village Fund is the second largest microcredit scheme in the world. RTG provided initial seed money of B 1 million (US$28,700) to each of Thailand's 74,000 villages and 4,500 urban communities. Hence, nearly 80,000 elected local Village Fund committees administer loans that reach 30 percent of all households (Boonperm et al. 2012).

In addition, the good leadership role played by the Bank of Thailand (BOT) in enhancing financial inclusion and actively implementing strategies, policies, and regulations to ensure an enabling environment was also instrumental in increasing access to finance. BOT emphasized its commitment to financial inclusion in its recent Financial Sector Master Plan (Phase II, for the period 2010–14), with the aim of increasing efficiency, promoting inclusion, and providing consumer protection, and it has entered into discussions with commercial banks concerning business models that might help reach the rural poor (BOT 2009).

BOT also is looking into extending services tailored to the needs of unbanked and underserved populations, while maintaining stability in the system. According to BOT's financial access survey for 2010, of the 878 districts in Thailand, 302 (34.4 percent) do not have commercial bank branches. However, when branches of specialized financial institutions (SFIs) are included, the number of districts with no branch service drops to 92 (10.5 percent). This is largely due to the strong rural presence of the Bank for Agriculture and Agricultural Cooperatives and the Government Savings Bank's primarily urban footprint. When Village Funds are included, the financial services footprint extends to virtually 100 percent of the country (ADB 2013a). Hence, the recent decision by BOT to extend its supervisory and regulatory mandate to SFIs is an important regulatory development because of their extensive coverage of rural and poor communities in providing a much-needed source of funds. Proper regulatory guidelines are needed to manage risks as well as to protect consumers.

A proactive and respected Thai Bankers' Association (TBA) ably supports BOT. TBA is a founding member of the Association of Southeast Asian Nations (ASEAN) Bankers Association and plays a key role in representing the banking community in discussions with BOT, the Ministry of Finance, and other government bodies in formulating and implementing key economic and financial policies. TBA also works closely with the Foreign Banks' Association on various banking issues, including improvements to the payment and settlement systems. TBA has been a valuable partner to BOT and its Payment Systems Committee in fostering efficient, safe, and sound infrastructure to enable access to financial and payment services by the Thai people and in supporting economic development.

Cash Access within Eight Minutes

The development of financial infrastructure that enhances outreach to a wider customer base is an integral part of Thailand's success story. Under guidance from BOT, banks in Thailand are expanding their infrastructure covering all four regions of the country (map 4.1).

Map 4.1 Distribution of Financial Institution Branches, Automated Machines, and EFTPOS Terminals in Thailand, by Region, 2013

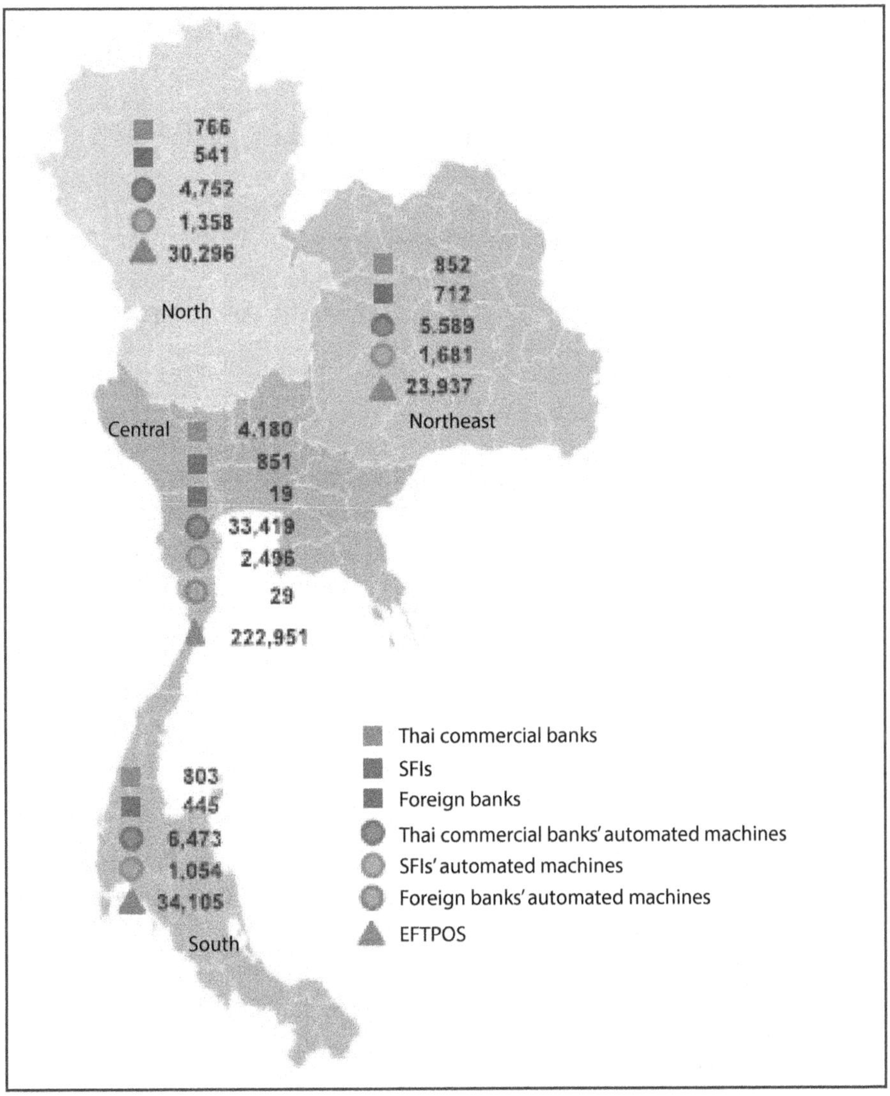

Source: BOT 2013. © Bank of Thailand. Reproduced, with permission, from Bank of Thailand; further permission required for reuse.
Note: EFTPOS = electronic funds transfer at point of sale; SFIs = specialized financial institutions.

Despite the economic contraction in 2013 that led to reductions in transaction volumes, mobile banking and Internet banking showed significant growth. The key payment infrastructure in the country comprises bank and nonbank branches, the Internet, mobile phones, ATMs, cash deposit machines (CDMs), and electronic funds transfer at point of sale (EFTPOS) terminals. This wide array of payment infrastructure ensures payment services to customers that are not restricted to bank and nonbank branch working hours.

In both 2012 and 2013, Thailand recorded significantly higher numbers of ATMs per 1 million persons (810 and 877 ATMs, respectively) than the average of Committee on Payments and Market Infrastructures (CPMI) countries (437 and 488 per 1 million persons, respectively).[23] Among Thailand's regions, ATM penetration indicates that, except for the Northeastern region (383 machines), the other three regions have higher ATM penetration per 1 million persons than the average for CPMI countries: Northern (524 machines), Southern (849 machines), and Central (1,977 machines).

However, the number of EFTPOS terminals per 1 million people (4,099 in 2013) was around half of the CPMI country average of 8,788 terminals (BOT 2013). In addition, although mobile-phone penetration was 142.7 percent, the number of mobile agreements was only around 1.16 million (1.8 percent of the population), indicating further opportunities for improvement (BOT 2013). In addition, Thailand's MNOs are already providing payment services and focusing on capturing the untapped market segments. The downside is that the proprietary mobile payment services and stand-alone systems used by each operator do not allow interoperability among MNOs, thus limiting uptake among users.

BOT is well aware that consumers in rural areas still face access as well as knowledge issues in terms of using e-money, and as a result, use the services of community financial institutions and informal sources at a higher cost.[24] The assessment on potential touchpoints by the 2013 FinScope Survey reveals that, while it takes an average of close to 36 minutes to reach a bank branch or ATM for the rural population, it takes only an average of 8 minutes to reach a grocery store (figure 4.8).

Hence, there is opportunity to provide financial services at a more affordable cost. BOT has issued agent banking regulations that allow for retailers to act as banking agents as well.[25] Already, Counter Service Co. Ltd. delivers payment services through more than 700 franchise holders. It has recently partnered with 7-Eleven stores and has access to their more than 8,300 outlets to deliver payment services.[26] 7-Eleven is now the most popular provider of utility and other payments in the country. Other providers include Jay Mart and Tesco Lotus. All these stores also offer point-of-sale (POS) devices and prepaid phone airtime replenishment ("top-up" services). Payment services are competitively priced and range from B 10 (around US$0.30) for utility bills to B 15 (around US$0.45) for payments due to some commercial banks and insurance companies (ADB 2013a). It is important to expand these services further to cover underserved areas and communities. The use of banking agents such as retail stores, as well as

Figure 4.8 Average Time to Financial Service Touchpoints in Thailand, 2013

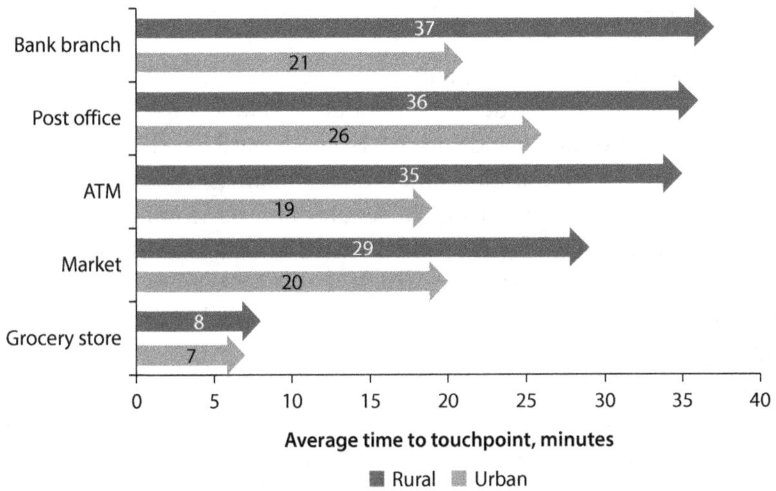

Source: ADB 2013a, based on FinScope Thailand 2013 Survey. © Asian Development Bank (ADB). Reproduced, with permission, from ADB; further permission required for reuse.
Note: ATM = automated teller machine.

mobile money solutions, would be the most cost-effective way to improve access to finance and payment services in these areas.

Banks also offer fund transfers competitively through their ATM networks (56,851 ATMs throughout Thailand in 2013 [BOT 2013]). ATM transfers are done free of charge within the same bank, while charges between two different banks cost around B 30 (about US$0.90). If the ATM of a different bank is used for withdrawals, up to four transactions per month are free of charge, while transactions beyond the fourth will have a B 10 fee. Moreover, Thailand Postal Company also acts as a banking agent to commercial banks and uses its 1,295 post offices to perform banking services. (Unlike retail shop payment agents, post offices are allowed to accept deposits.) The above analysis indicates that banks dominate the funds transfer and remittance markets.

An Interoperable, Multifunctional, Efficient ATM Network

Thailand is one of only three ASEAN member countries (along with Indonesia and the Philippines) that have full ATM interoperability, meaning that information can be exchanged across different technology systems and software applications (World Bank 2011). The fully interoperable ATM networks have expanded payment services such as bill and other payments as well as remittance transfers, and even airline e-ticketing and tax payments can be carried out at these stations. In addition, automated deposit machines (ADMs, also called CDMs) offer a cash deposit service, while passbook update machines (PUMs) provide convenience to customers without visiting bank branches. Some banks have unique three-in-one compact terminals featuring integrated note acceptance, dispensing, and self-service passbook update. The terminals' cash recycling capabilities also reduce operational costs for the banks.

The efficiency of the ATM network is also attributable to the National Interbank Transaction Management and Exchange (ITMX) Company, the interbank payment infrastructure provider.[27] ITMX supports all types of electronic payments and funds transfers from various channels, including ATMs, counters, Internet, phone, and mobile channels, and it fosters interoperability among banks by using a secure and efficient open platform. This eliminates the need for individual financial institutions to upgrade and maintain their own payment infrastructure systems. From the clients' perspective, interoperability has allowed them increased access through more bank-neutral touchpoints.

Another important element in Thailand's extensive ATM network is efficient cash management. With more and more banks focusing on providing ATM services to customers farther away from their head offices or even branches, it is important to manage timely cash replenishment and 24/7 service. In Thailand, specialized cash management companies such as G4S (a global integrated security company) manage ATMs on behalf of banks, including cash forecasting, cash transportation, reconciliation services, first-line maintenance, and ATM engineering services. All of these activities significantly contribute to the efficiency with which banks provide payment services to their clients.

In reviewing Thailand's financial inclusion strategy, it is apparent that although digital finance plays a significant role in the payment space, cash is still the dominant medium and seen as a free service by banking sector clients because of low transaction costs. Use of service providers by different customer groups shows the dominance of banks in the remittance and fund-transfer space (table 4.2). Efficiently managed ATM proliferation has given banks the leading edge in these areas.

Table 4.2 Financial Service Providers in Thailand, by Customer and Transaction Type, 2013

Customer type	*Loans*	*Savings*	*Utility payments*	*Remittances*
Farmer	BAAC	BAAC	7–11, door-to-door	ATM, postal orders
Poor villager	Village Fund, BAAC, moneylenders	BAAC, banks, informal or CDD savings group	7–11, door-to-door, utilities office	ATM, postal orders
Poor urban dweller	NBFI, moneylenders, pawnshops	Banks	7–11, trade outlets	ATM
Middle class	Banks, NBFI, credit union	Banks	7–11, trade outlets, banks	ATM, cashiers' checks
Businessman	Banks, NBFI	Banks	Banks	ATM, cashiers' checks

Source: ADB 2013b.
Note: ATM = automated teller machine; BAAC = Bank for Agriculture and Agricultural Cooperatives; CDD = Community Development Department; NBFI = nonbank financial institution; "Village Fund" refers to the Village and Urban Revolving Fund, a semiformal microcredit institution established by the government to make credit available to the villages.

The importance of proximity to touchpoints in utility payments is evident in the fact that retail chains such as 7-Eleven play a major role in this payment area. By regulation, savings and deposit markets are served only by banks and nonbank financial institutions such as SFIs. Looking at the market structure and the type of customers, it is evident that poor and underserved customers have limited access to digital money and payments, given that Village Funds, other community-based organizations, and cooperatives are cash-based. Hence, there is a huge opportunity for digital finance to reach the poor and the underserved, who are primarily being served by these institutions.

The Thailand case study reveals the impact of efficient coordination of policies and strategies toward payment services and access to finance through reduction of infrastructure costs by having single or near-single interoperable platforms. It also brings to light the government's and BOT's strategic involvement in these efforts.

Going forward, even though cash-based payments through ATMs and EFTPOS terminals are well coordinated in Thailand, cash is not the most efficient method of payments. Cash management imposes a high cost on governments, banks, and even businesses. There are many other payment modes that are more efficient and safer than cash, and moving toward e-money and digital finance should be Thailand's next goal. Cash has to be provided by BOT at a cost, and that does not help financial deepening. Cash-led financial systems find it difficult to graduate to the next level because of the retail focus of cash-based transactions.

From a macroeconomic point of view, Thailand has to come out of its middle-income trap by elevating small and medium enterprises (SMEs), microenterprises, and retailers to middle-income groups. E-money and e-commerce are the needs of the day. There is an opportunity for BOT to introduce new regulations and best practices for MNOs and other payment service providers to enter into the trade.

From a financial inclusion perspective, if mobile money services become interoperable and connect with the SFIs and also link with retail chains, the opportunity to reach the poorest by digital means will be limitless. These systems should encourage and incentivize the use of alternatives to cash. Furthermore, cost efficiencies would be significant in terms of reducing the high cost of cash management to the government as well as to other stakeholders. BOT appears to be aware of these benefits and plans to address these issues through its payment systems development strategy as well as the financial inclusion strategy.

The Philippines: The World's Oldest Mobile Money Initiative Has Yet to Reach Potential

The Philippines was one of the pioneers and earliest adopters of mobile money services, even before Kenya's M-Pesa became a household name. Launched in 2001, Smart Communication's SMART Money service was the first to market; Globe Telecom's GCash service followed in 2004. Despite being early movers in the game, mobile money hasn't scaled up to the levels initially predicted. By 2014, active mobile money users were still fewer than 7 million on a combined subscriber base of over 105 million mobile subscribers. In 2014, 69 percent of the population had neither an account in a formal financial institution nor a mobile money account—hence were excluded from the formal financial sector.[28]

The archipelagic structure of around 7,100 islands in the Philippines poses a big challenge to financial access, with most people living a great distance from financial services. Transporting money also has serious security concerns in the islands. As of 2014, more than 605 of 1,634 cities and municipalities did not have banks, and around 8 million people were living in these unserved areas (BSP 2015).[29] Conversely, almost 43 percent of all deposit accounts in the Philippines, amounting to 71 percent in value terms, were in the National Capital Region, commonly known as Metro Manila, with a population of just over 12 million of the Philippines' 100 million population (Tayag 2014).

At a glance, therefore, the Philippines seems to be the ideal place for mobile money to proliferate. Mobile-phone penetration is over 110 percent, and the country is called the texting capital of the world, with over a billion texts sent every day (GSMA 2014).[30] The Philippines has the eighth largest number of Facebook users in the world, and 36 percent of the population use the Internet. Furthermore, the average monthly remittances from overseas Filipinos to their beneficiaries in the Philippines have exceeded US$2 billion since 2013 (Martin 2017).[31] Moreover, the policy makers and regulators are flexible in enabling innovative money services to function. Hence the low mobile money user numbers continue to baffle the mobile money experts. While there can be many constraints for mobile money adoption and digital inclusion in the Philippines, this study examines the policy and regulatory aspects that help or hinder this process.

Regulators Creating an Enabling Environment

The Ministry of Finance and Bangko Sentral ng Pilipinas (BSP) have played an active role in creating an enabling environment to extend financial services to the poor and rural regions of the country. BSP adopted the "test-learn-regulate" model by allowing two nonbank institutions to offer e-money, and it issued progressive regulations and guidelines. In doing so, BSP aimed to increase financial outreach as well as to provide more options for people to receive international remittances[32] with the intention of reducing the cost of remittances through increased competition. By issuing Circulars Nos. 240 and 269 in 2000, BSP enabled e-money issuance.[33]

Consequently, SMART Money piloted and launched a bank-led model in partnership with Banco De Oro (BDO). Although Smart Communications manages the product and branding, the underlying prepaid "e-wallet," which is not considered to be a deposit, is held with BDO and is also linked to a SMART Money MasterCard debit card that can be used at ATM and POS terminals.[34]

In 2004, Globe Telecom launched a nonbank-led model, GCash, through a fully-owned subsidiary called G-Xchange Inc. by obtaining a license as a remittance service provider. The acceptance of money by a nonbank was deemed not to fall under the definition of banking, because no intermediary function was completed by the on-lending funds so received. GCash funds received by G-Xchange are held in partner commercial banks as wholesale deposits. These flexible but ad hoc arrangements that enabled operation of both types of e-money models were largely formalized under one regulation with the issuance of Circular No. 649 in 2009. This circular regulates both bank-led and nonbank-led models under one license for e-money issuers (EMIs).

To better understand the technological developments and innovative solutions offered by the service providers and the associated risks, in 2005 BSP established a specialized team of former bank examiners and information systems experts, the Core Information Technology Support Group (CITSG). CITSG on-site inspections follow COSO53[35] guidelines on internal controls and, among other aspects, examine logical access controls, record keeping, audit trail, disaster recovery, and business continuity planning. This group studies mobile financial services and other electronic financial services, their risks, and effective supervision methods. All specialists are certified in information systems auditing (Chatain et al. 2008). In 2006, guidelines were issued on technology risk management to ensure that banks have the knowledge and skills necessary to understand and effectively manage their technology-related risks.[36] In addition, guidelines for consumer protection for e-banking were issued in 2006.[37]

Following the 2001 placement of the Philippines on the FATF list of noncompliant countries and territories, both BSP and the Ministry of Finance worked diligently to make progress toward implementing plans to fight terrorism finance and money laundering. As a result, the Philippines was subsequently removed from the FATF watch list in 2005. The country demonstrated a strong commitment to implementing an AML/CFT regime in the context of its financial inclusion strategy. The AML regulation issued for mobile money includes the conduct of KYC and CDD procedures, record keeping, and reporting of suspicious transactions to the Anti-Money Laundering Council.

Regulatory Barriers that Hinder Scaling-Up

Although progressive policies adopted by BSP in finding innovative solutions to enable e-money operations by bank as well as nonbank EMIs resulted in successful e-money deployments of both types of models, scaling-up of these operations remains below expectations to date. In the absence of the Payment and Settlement law,[38] BSP proactively issued e-money regulations and operational guidelines under BSP's charter. Passage of the Payment and Settlement law will provide greater clarity on the regulatory mandate and enhance BSP's legal and enforcement authority.

In the absence of such broader powers, it appears that some of the regulations issued by BSP have had unintended consequences. Given the earlier blacklisting by FATF, BSP had issued detailed AML/CFT guidelines, which were very restrictive on the use of agents. According to Circular No. 471 of 2005, all agents have to be registered with the BSP as remittance agents by providing appropriate documentation and are subjected to the AML Act. Such agents also have to attend mandatory AML training seminars typically held in Manila.[39] Additionally, according to the guidelines, every time funds are transmitted or received (cash-in/cash-out), the agents are mandated to maintain originator information by requiring the sender to fill out personal information, recipient, and source of funds, among other things. Transaction records are supposed to be kept for five years. The ID check is carried out once, using one of the 20-plus ID types approved by the BSP.[40]

Furthermore, the agents cannot perform opening of new accounts. Such onerous rules are deterrents to the scaling-up of agent networks. As a result, the outreach of the MNOs has been limited, and people who want to use these services are discouraged. Authorities could consider allowing nonbank agents to open customer accounts, allowing remote registrations with lower thresholds, and allowing for tiered KYC procedures to encourage more agents and customers to participate in the mobile money schemes.

Mobile money services are trying to gain traction in the payment and funds transfer space against stronger and more established alternative channels, such as pawnshops for money transfer and banks or payment centers for bill payment. It is, therefore, crucial that mobile money providers offer a cheaper, less complicated solution to attract the lower-value segments. Hence, simpler regulatory guidelines would enable mobile money providers to compete in the low-income market by providing consumers with a low-cost and effective option, thereby addressing many of the Philippines' financial access issues.

Maldives: Mobile Money Opportunity Still Knocking at the Door

The Maldives case study[41] highlights why a unique solution developed to address a distinct financial access problem was not successful because of coordination issues at the macro level between the policy maker and the regulator. Today, however, changed dynamics may provide a much-needed solution, as forward-thinking policy makers and the regulators are trying to resolve the persistent problem through a coordinated approach that would allow for a low-cost, innovative digital solution.

The Special Financial Inclusion Challenge in Maldives

Maldives is a challenging environment to deliver access to financial services through traditional bank branch networks. The country spans 1,190 low-lying coral islands that stretch over 90,000 square kilometers of Indian Ocean, 99.5 percent of which is water. Nearly 338,400 people live on 200 scattered inhabited islands in 26 atolls, and one-third of the inhabited islands have a population of less than 500, while 71 percent of the inhabited islands have fewer than 1,000 people (MMA 2015). Although the number of bank accounts exceeded the population by 2014, people living in the outer islands and atolls experience difficulties in accessing financial products and services through traditional banking outlets and existing financial infrastructure such as bank branches, Internet banking, ATMs, and POS terminals.

The wide geographical dispersion and lack of financial institutions and infrastructure translates into high transaction costs, decreased financial accessibility, and greater security risks for people living and working in the atolls. For example, people reportedly often send their ATM cards along with their Social Security and personal identification numbers (PINs) through their friends or relatives to withdraw cash (pension money, remittances, and other receivables) from ATMs located in Malé or other islands where ATM services are available. Thus, cash remains the dominant payment instrument in the islands and outer atolls. Table 4.3 illustrates the increase in cash in circulation from 2007 to 2014.

Bank of Maldives (BML) is the largest commercial bank, with 27 branches (of which 20 are in atolls), 8 mobile Dhoni units,[42] and 68 active ATMs (of which 27 are located in atolls). As of April 2014, the government of Maldives owned 62 percent of the shares of BML. The Bank recently launched self-service banking through 22 multifunctional ATMs installed across Malé and introduced its first U.S. dollar ATM at BML headquarters in Malé. To facilitate payment requirements of outer islands and atolls, BML operates an agent-based banking solution backed by BML banking agents in 70 islands through Internet banking and 3,610 POS card terminals (of which 1,087 POS merchants are in atolls including resorts). BML provides clearing and settlements through its own switch. According to BML, there is now a branch presence in every atoll in the country except one.

From BML's point of view, it is neither economical nor scalable to provide agent-based banking services to all islands and atolls. In a rudimentary form, BML operates cash-back POS facilities as part of its agent-based banking services. By 2016, BML had announced plans to facilitate full person-to-person (P2P) transactions through app-based mobile-phone banking facilities (already started) and real-time POS transactions. This commitment was not fully realized.

Table 4.3 Cash in Circulation in Maldives, 2007–14

Year end	*Rf (millions)*
December 2007	1,049.0
December 2012	2,475.5
December 2013	3,252.4
December 2014	3,099.4

Source: MMA 2015, table 7.1.
Note: Rf = Maldivian rufiyaa.

Meanwhile, many of the islands are still being served by BML's Dhoni cash distribution system and limited agent-based banking facilities. However, anecdotal evidence suggests that Dhoni banking is unreliable in some atolls, and the payment services offered by BML are unaffordable, even for those who live in Malé, because of relatively high transaction costs—a consequence of BML's dominant position. For example, BML's own card customers are required to pay 2.5 percent of the total value of a transaction for using BML's own POS machines, which is relatively high compared with other countries in the South Asia region. Hence, BML's branch network, agent network, and infrastructure, although expanding, still leave many smaller outlying islands unserved.

On the other hand, operations of the other commercial banks (established as branches of foreign banks), with the exception of Maldives Islamic Bank Pvt. Ltd., are largely confined to Malé and mostly serve the corporate clients. Although Maldives Islamic Bank shows interest in community banking, other commercial banks have little interest in developing banking services in the outer atolls. A few nonbank institutions also operate POS acquisition services in resorts. Hence, in terms of availability and affordability of financial products and services, the solutions to date seem to have fallen short in providing adequate financial and payment services.

Seeking a Unique Solution to a Distinct Problem

To enhance access to finance and provide convenient payment facilities to people in the outer islands and atolls, in 2001, CGAP and the World Bank together designed and funded the Mobile Phone Banking Project (World Bank 2015). The objective of the project was to build an integrated gross and retail national payment system with a first-of-its-kind interoperable payment switch that connects an interoperable mobile payment system (MPS, also named Keesa)[43] that would allow all cooperating Maldives-based banks and all mobile telephone operators to share the payment infrastructure of the country and take advantage of the extensive infrastructure of small shops on all inhabited islands, which could serve as cash-handling points on behalf of banks and MNOs. The Keesa system would have allowed subscribers to make payments from and receive payments to these accounts via mobile handsets and the Internet very conveniently. The system was installed at the Maldives Monetary Authority (MMA), which completed the initial user acceptance testing of the system in August 2011. The design and commencement of the MPS was based on the memorandum of understanding (MOU) signed between the banks and MMA.

Despite the technical completion of the system and willingness of most banks to serve their customers through Keesa, the project could not be operationalized because BML pulled out of the project during the last stages, citing business reasons. Given that BML accounts for more than 60 percent of the market share of customers and has 27 branches, its participation was crucial for the success of Keesa. Many attempts were made by MMA to reassure BML and accommodate its requests. MMA brought in added functionalities at the technical level to minimize the cost of interfacing for the participating banks, such as various reporting systems and a "Converter Gateway" to facilitate payments. However, BML refused to participate despite the MOU.

Unfortunately, the regulator (MMA) was unable to persuade BML to participate in a national system that was designed to help provide access to a large segment of the society, and the major shareholder (Ministry of Finance and Treasury) was unwilling to interfere in BML's business decision. The project, therefore, was closed in 2013, having only operationalized the real-time gross settlement (RTGS) system and part of the automated clearinghouse (ACH) and unable to operationalize the mobile banking and the interoperable payment switch. As a result, the outer islands and atolls still have limited financial access.

Positive Macro Dynamics: Collaboration between Policy Makers and the Regulators

Having realized the practical difficulties in moving toward a national payments switch in the short term given the market dynamics, MMA has now decided to establish a nonbank-led mobile money solution to facilitate financial access in the short term. Since MNOs are nonbank institutions that are not under the direct supervision of MMA, it is necessary for MMA to legally allow them to enter the payment space by issuing regulations governing the nonbank-led mobile solution to manage any operational risks as well as to establish public confidence in the proposed system.

MMA is keen to avoid overregulation and desires an interoperable, simple, and moderately regulated solution with low maintenance costs. This idea was strongly endorsed by the Ministry of Finance and Treasury. The Communications Authority of Maldives also strongly endorsed an interoperable nonbank-led mobile-money solution where both MNOs participate to maximize the value proposition to the customers.[44]

This uniquely collaborative policy-regulatory drive had resulted in both MNOs showing a strong interest in launching mobile money solutions. Therefore, the government and MMA once again sought technical support from the World Bank in their efforts to establish an enabling environment for a nonbank-led mobile money solution aimed at enhancing access to financial and payment services in the Maldives, especially in the outer islands and atolls, and to ground this in a well-designed legal and regulatory framework. The World Bank advised the MMA and the government on this process by way of a Policy Note, and provided the technical support for this important initiative.

Through two advisory engagements, the World Bank supported MMA and the government. The first effort produced a May 2015 Policy Note entitled "Maldives: A Mobile Money Operator-Based Mobile Money Solution," which provided a road map for implementing an MNO-led mobile money solution (M-Wallet). On September 25, 2015, the World Bank's FIRST Initiative approved a three-year grant-funded technical assistance project entitled

"Maldives: Enabling a Non-Bank Mobile Money Solution." The grant has provided the source of funds for expert advisory assistance to MMA, the government, and Maldives' MNOs to help implement (according to international best practice) custodian and trustee arrangements, risk mitigation and consumer protection, legal and regulatory amendments, and payment system oversight and coordination mechanisms.

Progress is slowly but surely being made. The enabling regulation for MNO-led mobile money solutions has been provided for under the Maldives Monetary Authority Act while Maldives' Payment Law still awaits enactment in Parliament. With enabling regulation in place, both MNOs have activated e-wallets. While interoperability and other synergies are still to come, the country is committed to learning from the lessons of international good practice and to providing greater levels of financial access to its citizens.

Notes

1. Global Findex Survey 2014 data, http://datatopics.worldbank.org/financialinclusion/.
2. Consult Hyperion was the lead information technology (IT) consultancy company that developed the M-Pesa software.
3. Kenya financial inclusion data from Global Findex Survey 2014, http://datatopics.worldbank.org/financialinclusion/.
4. The retail payments system includes automated clearinghouse (ACH) transactions, automated teller machine (ATM) switches (Kenswitch and PesaPoint), over-the-counter remittances, credit or debit card and point-of-sale systems, and e-money schemes (mobile money and virtual money).
5. The "National Payment System Regulations, 2014" issued August 1, 2014, under the National Payment System Act (No. 39 of 2011).
6. "Banking Correspondent" refers to a financial model in India whereby a representative is authorized to offer services such as cash transactions where the lender does not have a branch.
7. "Self-help group," in this context, refers to a village-based financial intermediary committee that is most common in India, usually composed of 10–20 women.
8. Microinsurance is the protection of low-income people (those living on between approximately US$1 and US$4 per day) against specific perils in exchange for regular premium payment proportionate to the likelihood and cost of the risks involved. This definition is the same as one might use for regular insurance except for the clearly prescribed target market: low-income people. The target population typically consists of persons ignored by mainstream commercial and social insurance schemes, as well as persons who have not previously had access to appropriate insurance products.
9. Aadhaar unique ID and its link to DBTs are discussed in chapter 6 as a critical enabler. For more information about the unique ID program, see the Unique Identification Authority of India's website: http://uidai.gov.in/.
10. CDMA (Code Division Multiple Access) and GSM (Global System for Mobiles) are shorthand for the two major radio systems used in cell phones.

11. Global Findex Survey 2011 data, http://datatopics.worldbank.org/financialinclusion/region/south-asia/.
12. Payment and Settlement Systems Act, No. 28 of 2005; Convention on the Suppression of Terrorist Financing Act, No. 25 of 2005; Financial Transactions Reporting Act, No. 6 of 2006; Prevention of Money Laundering Act, No. 5 of 2006; Electronic Transactions Act, No. 19 of 2006; Payments Devices Frauds Act, No. 30 of 2006; Computer Crimes Act, No. 24 of 2007; and Secured Transactions Act, No. 49 of 2009.
13. The Local Treasury Bills Ordinance, No. 8 of 1923; Bills of Exchange Ordinance, No. 25 of 1927; Registered Stocks and Securities Ordinance, No. 7 of 1937; Banking Act, No. 30 of 1988; Evidence (Special Provisions) Ordinance, No. 14 of 1995; Consumer Affairs Authority Act, No. 9 of 2003; Intellectual Property Act, No. 36 of 2003.
14. Convention on the Suppression of Terrorist Financing Act, No. 25 of 2005; Financial Transactions Reporting Act, No. 6 of 2006; and Prevention of Money Laundering Act, No. 5 of 2006.
15. The Budapest Convention on Cybercrime is also known as the Council of Europe Convention on Cybercrime. It is the only available international treaty on the subject seeking to address Internet and computer crime by harmonizing national laws, improving investigative techniques, and increasing cooperation among nations.
16. LankaClear is owned by the CBSL and commercial banks.
17. National Development Bank partnered with Dialog Axiata, to offer the eZ Pay mobile money service.
18. Mobile Payments Guidelines, No. 1 of 2011, for the Bank-led Mobile Payment Services; and Mobile Payments Guidelines, No. 2 of 2011, for Custodian Account Based Mobile Payment Services. The entry requirements stipulated in the "Service Providers of Payment Cards Regulations, No. 1 of 2009" were further clarified in the "Payment Cards and Mobile Payment Systems Regulations, No. 1 of 2013."
19. Chinese mobile payments in general reached 1.7 billion transactions in 2013, up 213 percent from the previous year, according to figures the bank released earlier in 2013. Those transactions were worth US$1.6 trillion, up some 317 percent from 2012. Alibaba and Tencent are responsible for the bulk of those figures (Hernandez 2014).
20. "Legal certainty" is a principle that the law must provide those subject to it with the ability to regulate their conduct. Legal certainty is internationally recognized as a central requirement for the rule of law.
21. The Payment Cards and Mobile Payment Systems Regulation (No. 1 of 2013) replaces the Service Providers of Payment Cards Regulation (No. 1 of 2009).
22. Although the Global Findex 2014 Survey shows 78 percent of the adult population having access to formal financial institutions, there are definitional differences between FinScope and Findex.
23. Data from Committee on Payments and Market Infrastructures (CPMI) 2016 "Statistics on Payment, Clearing, and Settlement Systems in the CPMI Countries," http://www.bis.org/cpmi/publ/d155.pdf. Formerly known as the Committee on Payment and Settlement Systems (CPSS), established by the Bank for International Settlements (BIS), CPMI monitors payment systems developments in its 23 member economies annually. CPMI member economies include Australia; Belgium; Brazil; Canada; China; France; Germany; Hong Kong SAR, China;

India; Italy; Japan; the Republic of Korea; Mexico; the Netherlands; the Russian Federation; Saudi Arabia; Singapore; South Africa; Sweden; Switzerland; Turkey; the United Kingdom; and the United States.

24. These costs include higher direct costs such as fees as well as indirect costs such as transport costs, loss of daily wages, and so on.
25. Bank of Thailand Notification No. SorNorSor. 9/2553: Guideline for Appointing Banking Agents. (In terms of the guidelines, agent activities are restricted depending on the type.)
26. By 2016, there were more than 9,542 7-Eleven stores throughout Thailand, 44 percent of them in the Bangkok area. For more about CP All Public Company Ltd., see "CP All," https://en.wikipedia.org/wiki/CP_All.
27. ITMX was founded by the Thai Bankers' Association as the ATM Pool Co. Ltd. in 1993, renamed as National ITMX in 2005. Created to satisfy Thailand's requirement to keep up with continuing global advancement in electronic commerce and payment systems, its business policy follows the BOT's Payment Systems Roadmap 2004.
28. Financial inclusion data from the 2011 and 2014 Global Findex Surveys, http://datatopics.worldbank.org/financialinclusion/.
29. The Central Bank of the Philippines' (Bangko Sentral ng Pilipinas, or BSP) National Strategy for Financial Inclusion states that the presence of other financial service providers such as pawnshops, remittance agents, money changers or foreign exchange dealers, e-payment service providers, mobile banking agents, nonstock savings and loan associations (NSSLAs), and credit cooperatives have helped significantly to enhance the access to financial products and services in the unbanked areas. Accordingly, this relates to 50,000 touchpoints, reducing the unserved municipalities from 36 percent to 12 percent (196 municipalities); however, the number of adults with accounts remains unchanged (BSP 2015).
30. However, unique mobile subscribers were 50.9 million in 2014, with an average of two SIMs per subscriber (GSMA 2014).
31. In 2014, 38 percent of Filipino adults were receiving remittances from a family member abroad. There were more than 10 million overseas Filipinos in 2013. Remittances reached an equivalent of 8.5 percent of the country's GDP in 2014 (BSP 2015).
32. Overseas Filipinos send an average of US$2 billion per month in international remittances, equivalent to 8.5 percent of GDP (BSP 2015).
33. Circular No. 240 of 2000 required all banks to obtain prior approval from BSP before launching e-banking services, while Circular No. 269 of 2000 set out the approval process for a new e-banking service.
34. The SMART Money service earns a commission on the bank interchange fee charged to merchants using banks other than BDO.
35. The Committee of Sponsoring Organizations of the Treadway Commission (COSO) is a nonprofit organization providing thought leadership and guidance on internal control, enterprise risk management, and fraud deterrence.
36. Circular No. 511 of 2006.
37. Circular No. 542 of 2006.
38. A draft Payment and Settlement bill is in the parliament.
39. The BSP has agreed to allow EMIs to train their own agents and now allows them to register many remittance agents with a single application. However, regulations are yet to reflect these amendments.

40. There is no national ID in the Philippines, and the impact of having 20+ types of ID documents is discussed in chapter 6, "Unique Identification."
41. This case study was informed by the Policy Note entitled "Maldives: A Mobile Money Operator-Based Mobile Money Solution," prepared by Anoma Kulathunga and Ranee Jayamaha (South Asia Finance and Markets Global Practice) as one of two outputs from a World Bank mission to Maldives March 1–13, 2015. The mission team was led by Thyra Riley.
42. Bank of Maldives PLC (BML), as Maldives' largest bank, provides financial services throughout the country's hundreds of atoll islands through 177 cash agents, 12 Self-Service Banking Centers, and 5 Dhoni Banking Units. "Dhoni" refers to a traditional multipurpose boat used in the Maldives. Dhoni Banking Units—comprising BML staff teams—make over 2,000 trips a year to the distant atolls and outer atoll islands to pay salaries and pensions, and to provide cash management services to agents and ATMs.
43. "Keesa" is Dhivehi for "wallet."
44. Maldives has two MNOs: Dhiraagu and Ooredoo.

Bibliography

ADB (Asian Development Bank). 2013a. "Kingdom of Thailand: Development of a Strategic Framework for Financial Inclusion. Thailand Financial Inclusion Synthesis Assessment Report." Project No. 45128, ADB, Manila.

———. 2013b. "Kingdom of Thailand: Development of a Strategic Framework for Financial Inclusion. Thailand Microfinance Supply-Side Assessment Report." Technical Assistance (TA) 7998, ADB, Manila.

BIS (Bank for International Settlements). 2012. *Basel Core Principles for Effective Banking Supervision*. Basel: BIS.

Boonperm, J., J. Haughton, S. Khandker, and P. Rukumnuaykit. 2012. "Appraising the Thailand Village Fund." Policy Research Working Paper 5998, World Bank, Washington, DC.

BOT (Bank of Thailand). 2009. "The Financial Sector Master Plan Phase II." Executive summary, BOT, Bangkok.

———. 2013. *Payment Systems Report 2013*. Bangkok: BOT.

BSP (Central Bank of Philippines, Bangko Sentral ng Pilipinas). 2015. "National Strategy for Financial Inclusion." Booklet, BSP, Manila.

CBSL (Central Bank of Sri Lanka). 2014. *Annual Report*. Colombo: CBSL.

CGAP (Consultative Group to Assist the Poor). 2013. "Direct Benefit Transfer and Financial Inclusion: Learning from Andhra Pradesh." Research results summary, CGAP, Washington, DC.

Chakrabarty, K. C. 2012. "The First Mile Walk into the Financial System." Address by the deputy governor, Reserve Bank of India, at the Financial Inclusion Conference 2012, New Delhi, August 7.

Chatain, P., R. Hernández-Coss, K. Borowik, and A. Zerzan. 2008. "Integrity in Mobile Phone Financial Services: Measures for Mitigating Risks from Money Laundering and Terrorist Financing." Working Paper 146, World Bank, Washington, DC.

Cirasino, M., and J. A. Garcia. 2008. "Measuring Payment System Development." Working Paper 49003, World Bank, Washington, DC.

Cook, T. 2015. "An Overview of M-PESA." Online FSD Kenya article, August 12. http://fsdkenya.org/an-overview-of-m-pesa/.

CRISIL. 2013. "CRISIL Inclusix–Vol. 1." Key findings from first Inclusix index, CRISIL, Mumbai. https://www.crisil.com/about-crisil/crisil-inclusix.html.

Daily FT. 2015a. "North Beats Western Province in Banking Density." January 6.

———. 2015b. "Sri Lanka Ratifies UN Electronic Communications Convention, Another First for South Asia." July 17.

Debu C. 2015. "MUDRA Bank: Weighing the Possible Benefits." MapsofIndia.com article, April 8.

di Castri, S. 2013. "Enabling Mobile Money Policies in Sri Lanka: The Rise of eZ Cash." Case study, Mobile Money for the Unbanked Program, Groupe Speciale Mobile Association (GSMA), London.

Economic Times. 2014. "PM 'Jan Dhan' Yojana Launched; 1.5 Crore Bank Accounts Opened in a Day." *IndiaTimes: The Economic Times*, August 29.

Fernando, J. 2013. "E-transactions to M-transactions: Serving the Next Generation Customers." Paper presented at the 25th Anniversary Convention, Association of Professional Bankers, Colombo, Sri Lanka. October 8–9. http://www.apbsrilanka.org/articales/25_ann_2013/2013_12_Jayantha%20Fernando.pdf.

FinScope. 2016. "FinScope Consumer Survey Thailand 2013: Survey Highlights." Booklet, FinMark Trust, Johannesburg, South Africa.

FSD (Financial Sector Deepening) Kenya. 2007. "Financial Access in Kenya: Results of the 2006 National Survey." Survey results report, FSD Kenya, Nairobi.

———. 2013. "FinAccess National Survey 2013: Profiling Developments in Financial Access and Usage in Kenya." Survey results report, FSD Kenya, Nairobi.

GSMA (Groupe Speciale Mobile Association). 2014. "Country Overview: Philippines Growth through Innovation." Analysis, GSMA, London.

———. 2015. *The Mobile Economy 2015*. Annual global report. London: GSMA.

Hernandez, Will. 2014. "Is China's QR-Code Ban about Security or Lost Revenue?" MobilePaymentsToday.com, March 24.

IMF (International Monetary Fund). 2015. "Thailand—2015 Article IV Consultation—Staff Report." Country Report No. 15/114, IMF, Washington, DC.

Jayamaha, R. 2014. *The Money Pipeline: A Pillar of Financial Stability*. Sri Lanka: privately printed.

Kulathunga, A., and R. Jayamaha. 2015. "Maldives: A Mobile Money Operator-Based Mobile Money Solution." Policy Note, World Bank, Washington, DC.

Kumar, K., and D. Radcliffe. 2015. "2015 Set to Be Big Year for Digital Financial Inclusion in India." Consultative Group to Assist the Poor (CGAP) blog, January 15. http://www.cgap.org/blog/2015-set-be-big-year-digital-financial-inclusion-india.

Leonard, M., J. Dahiya, T. V. S. Ravi Kumar, I. Wijesiriwardana, C. Linder, and G. A. N. Wright. 2011. "Deposit Assessment in Sri Lanka: Industry Mapping of Small Balance Deposits in South Asia." Working Paper 94762, World Bank, Washington, DC.

Martin, K. 2017. "OFW Remittances Up to 6.2 Percent." *Philippine Star*, February 17.

MMA (Maldives Monetary Authority). 2015. *Monthly Statistics* 16 (7). MMA, Malé.

MMAI and GSMA (Mobile Money Association of India and Groupe Speciale Mobile Association). 2013. "Mobile Money: The Opportunity for India." Position Paper 13,

Submission to the Reserve Bank of India's Committee on Comprehensive Financial Services for Small Businesses and Low-Income Households, GSMA, London.

Patel, A. 2016. "Financing Small Farmers for India's Food Security." *International Journal of Research–Granthaalaya* 4 (7): 196–212.

Peake, C. 2012. "New Frontiers: Launching Digital Financial Services in Rural Areas." Policy brief presented at the Ninth Annual Brookings Blum Roundtable on Global Poverty, Aspen, CO, August 1–3.

PMJDY (Pradhan Mantri Jan Dhan Yojana). 2015. "No. of Accounts Opened under PMJDY as on 27.05.2015 (Summary)." Progress report, PMJDY, Ministry of Finance, Government of India, New Delhi. https://pmjdy.gov.in/ArchiveFile/2015/5/27.05.2015.pdf.

Rao, P. S. M. 2015. "Heavy Inflow, but Farm Credit Seldom Reaches Fields." *Deccan Herald*, June 20.

RBI (Reserve Bank of India). 2014. "RBI Releases Report of the Committee on Comprehensive Financial Services for Small Businesses and Low-Income Households." Press release, January 7.

Sivaiah, K., and V. B. Naidu. 2015. "Growth and Structure of the Cooperative Agricultural Credit System in India." *International Journal of Multidisciplinary Research and Development* 2 (2): 292–95.

Tayag, P. B. R. 2014. "Strengthening Financial Inclusion through an Enabling Policy and Regulatory Environment." PowerPoint presentation by the head of Inclusive Finance Advocacy, Central Bank of the Philippines, to "Expert Meeting on the Impact of Access to Financial Services, Including by Highlighting Remittances on Development: Economic Empowerment of Women and Youth," United Nations Conference on Trade and Development (UNCTAD), Geneva, November 12–14.

TRAI (Telecom Regulatory Authority of India). 2013. "Highlights on Telecom Subscription Data as on 30th September, 2013." Press Release No. 78/2013, November 5.

Wharton School. 2014. "Financial Inclusion in India: Moving Beyond Bank Accounts." Knowledge@Wharton online journal, September 18. Wharton School, University of Pennsylvania, Philadelphia.

World Bank. 2008. "Payment Systems Worldwide: A Snapshot. Outcomes of the Global Payments System Survey 2008." Financial sector study, World Bank, Washington, DC.

———. 2011. "Payment Systems Worldwide Global Payment Survey 2010: Country-by-Country Appendix." World Bank, Washington, DC.

———. 2015. "Maldives Mobile Phone Banking Project." Implementation Completion Report Review (ICRR) No. 14559, World Bank, Washington, DC.

CHAPTER 5

Innovative Uses of Infrastructure and Digital Ecosystems

Introduction

The development of digital services offers a huge opportunity to advance financial inclusion and develop new markets through business models that are viable for the base of the pyramid. A critical first step is to ascertain that a conducive digital ecosystem and infrastructure are in place that can reach down to the poor and vulnerable. For innovative digital financial initiatives to become reliable conduits for financial inclusion, they must be accessible quickly and efficiently to promote fast, reliable, simple, and affordable solutions. A major challenge is putting in place all of the elements needed to go the last mile in terms of connectivity.

This chapter focuses on the critical meso- and micro-level elements of interoperability; agent network management; mobile money add-on applications; and biometric-enabled, card-based grant payment disbursement systems. The analysis highlights the efficiency and proactive use of each of these elements by the case study countries in delivering results and reaching meaningful scale. In terms of financial inclusion, scaling-up does not mean just commercial viability; it implies reaching the unbanked masses in a sustainable manner. The case studies discussed in this chapter have reached or have shown the potential to reach scale and have demonstrated how the ecosystem and infrastructure can be used as a digital rail system to deliver an array of affordable services to the poor and the vulnerable.

Interoperability in Indonesia, Pakistan, Sri Lanka, Tanzania, and Thailand

> *Competitive market conditions should be fostered in the retail payments industry, with an appropriate balance between cooperation and competition to foster, among other things, the proper level of interoperability in the retail payment infrastructure.*
>
> —World Bank, "Developing a Comprehensive National Retail Payments Strategy"

Results from the World Bank survey on innovations in retail payment instruments and methods highlight that the majority of innovative products and mechanisms have limited interoperability, meaning that information cannot flow seamlessly between different operating systems and platforms (World Bank 2012a). Less than 20 percent of the products were reported to be fully or partially interoperable. Around 25 percent of the products and services supported some mechanism to exchange funds with traditional payment products.[1] Although interoperability is considered to be best practice, evidently it is the exception rather than the norm.

Most digital financial systems launched are closed-loop systems and proprietary. Hence, the ability for the customers to transact with another service using a different provider's platform is limited. On the other hand, enhancing the value proposition to the customers and furthering the social agenda through interoperability may have costs in terms of possible higher rates,[2] lower levels of accountability, and lower levels of investment resulting from greater difficulty in capturing revenues relative to proprietary or closed-loop systems. Hence, it is necessary to evaluate what level of interoperability a country should aim for based on the financial landscape, the desired or targeted level of deepening, and the strategic policy agenda. Ideally, payment system interoperability involves the ability of the various players such as banks, nonbank financial institutions, payment system providers, mobile network operators (MNOs), card acquirers, governments, businesses, and customers to send, receive, and process funds, documents, and other instruments electronically through a common channel.

"Interoperability" is defined as "a situation in which payment instruments belonging to a given scheme may be used in platforms developed by other schemes, including in different countries. Interoperability requires technical compatibility between systems, but can only take effect where commercial agreements have been concluded between the schemes concerned" (World Bank 2012a). In the context of retail payment, multiple levels of interoperability are identified: systemwide, cross-system, and infrastructure-level. A system that has only systemwide interoperability enables competition among the participants of that system; a system that has cross-system interoperability enables competition between systems; and a system that has infrastructure-level interoperability enables the same infrastructure to be used to support multiple payment mechanisms.

Given the focus on mobile money as a digital means of reaching the underserved, the study examines interoperability as one of the critical elements in enhancing financial and payment access to the poor. When a country has more than one MNO offering their own mobile money solutions on proprietary platforms with their own operating rules and networks of agents, measures to achieve interoperability among MNOs are likely to be needed to optimize outreach to customers who are not served or who are underserved by the banking community and to benefit from economies of scale, as well as to discourage anticompetitive practices.

In reality, mobile networks are interoperable at a technical level, just as banks are. Banks are able to allow customers of another bank to use their automated teller machines (ATMs) at a higher fee, as do mobile networks and service providers for calls between them. Although interoperability may exist at this technical level, at a functional or more practical level, such limited interconnectivity is not optimal.

In terms of mobile money interoperability, the important types are platform-level interoperability (which permits customers of one service to send money to customers of another service); agent-level interoperability (which permits agents of one service to serve customers of another service); and customer-level interoperability (which permits customers to access their account through any subscriber identification module [SIM]).

The most successful mobile operation, Kenya's M-Pesa, is not a model for interoperability as it is a closed-loop system. If an M-Pesa account holder wants to send money to a non-Safaricom phone, they can do so and M-Pesa will generate a one-time code for them. The recipient will have to take this code to an M-Pesa agent and must cash out; it is not possible to add to his or her e-wallet. Hence, interoperability was not a factor in M-Pesa's success.

Although M-Pesa's market dominance acted as a barrier to interoperability, interestingly, the recently passed Kenya National Payment System Regulations 2014 require "open systems capable of becoming interoperable with other payment systems in the country and internationally." This newly minted provision in the regulations helped the Kenyan Competition Authority to end agent exclusivity, so that agents are able to offer their services to any MNO.

Countries such as Indonesia, Pakistan, Sri Lanka, Tanzania, and Thailand have launched mobile money initiatives with diverse types of interoperability. Experiences from these countries are summarized in the subsections that follow.

Sri Lanka: World's First Full End-to-End Interoperable Mobile Money System

When Dialog Axiata, Sri Lanka's leading MNO, launched Sri Lanka's pioneering mobile money service, eZ Cash, in June 2012, Sri Lanka already had high inclusion numbers, with banks and nonbank financial institutions serving all parts of the island. The Global Findex Survey 2014 shows that 83 percent of adults had formal accounts in 2014, up from 69 percent in 2011.[3] In 2014, the share using financial institutions for saving and borrowing were 31 percent and 18 percent, respectively. Genderwise, Sri Lanka has the highest equality in the South Asia region, with 83 percent of adult women having accounts. Furthermore, the poorest also show very high inclusion numbers at 80 percent, while adults in rural areas recorded a high 83 percent (table 5.1).

From the beginning, Dialog Axiata understood that mobile money would be one more option for the Sri Lankans, and that they would not likely be able to have an exclusive large footprint. They further realized that to compete with the already established financial sector, the products and services that Dialog offered would have to be tailored to the customers and provide more in terms of cost-effectiveness, convenience, and value. Because not many Sri Lankans use credit cards or similar card instruments, Dialog championed the small but voluminous transactions that the banking sector was unlikely to pursue.

Table 5.1 Financial Inclusion in Sri Lanka Relative to South Asia and Lower-Middle-Income Countries, 2014
Percentage

Sri Lanka: Population, ages 15+ years: 15.3 million GNI per capita: US$3,170			
Survey item	*Sri Lanka*	*South Asia*	*Lower-middle-income countries*
Account (ages 15+ years)			
All adults	82.7	46.4	42.7
Women	83.1	37.4	36.3
Adults belonging to the poorest 40%	79.8	38.1	33.2
Young adults (ages 15–24)	85.2	36.7	34.7
Adults living in rural areas	83.4	43.5	40.0
Financial institution account (ages 15+ years)			
All adults, 2014	82.7	45.5	41.8
All adults, 2011	68.5	32.3	28.7
Mobile account (ages 15+ years)			
All adults	0.1	2.6	2.5

Source: Global Findex Survey 2014, http://datatopics.worldbank.org/financialinclusion/.
Note: GNI = gross national income.

In an unprecedented move, Dialog invited other MNOs to share the eZ Cash platform, with the wallet to be jointly held and managed. In the Sri Lankan market at this stage, only Dialog and Sri Lanka Telecom Mobitel had e-wallets. In 2014, Etisalat (the third largest MNO) and a few months later Hutch (the fourth largest MNO) also joined (figure 5.1). Hence, eZ Cash has the distinction of being the first mobile money system in the world to be end-to-end interoperable across multiple service providers. Dialog, Etisalat, and Hutch operate eZ Cash as a single wallet.

Today, the eZ Cash client base has grown to over 2.8 million Sri Lankan mobile users who are connected with 20,000 merchants and service providers through the country's largest mobile payment and transaction gateway. The combined subscriber strength of the three MNOs—over 15 million, or more than 65 percent of the entire Sri Lankan population—now have instant seamless access to the full portfolio of eZ Cash services, and can transact electronically via eZ Cash.

The Fully Interoperable Model

Dialog has enabled access to eZ Cash services via an unstructured supplementary service data (USSD) menu by typing #111# on mobile phones. Etisalat and Hutch subscribers who register for the eZ Cash service use the same USSD menu. Even the costs (service charges) are the same (table 5.2). Interestingly, Mobitel (the second largest MNO in Sri Lanka), which launched a separate mCash wallet, has the same low fees.

The ability to simply dial in and self-register for the basic wallet without going to an outlet or kiosk is hugely convenient. Dialog currently has over 20,000

Figure 5.1 Market Share of Sri Lankan Mobile Service Providers, 2014

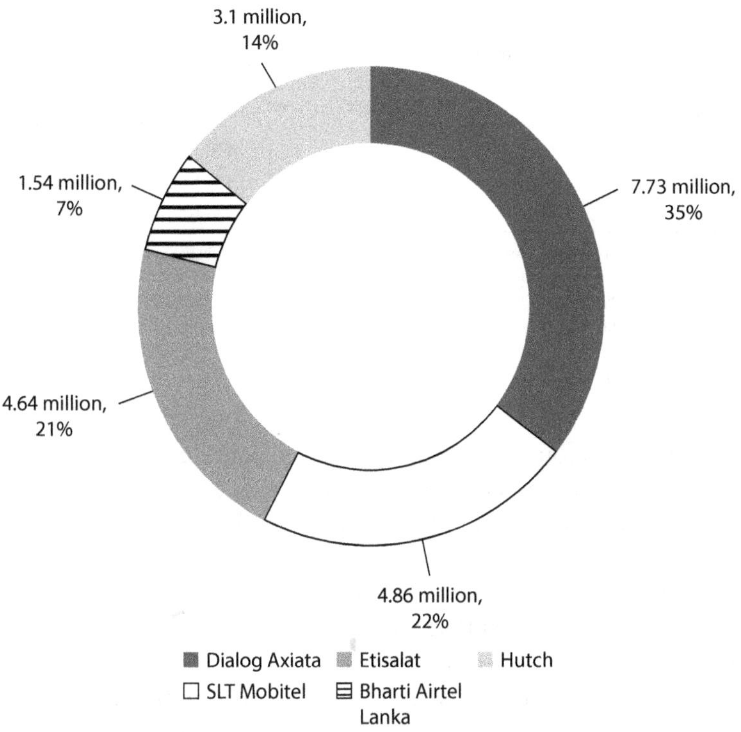

Source: diGIT IT 2014.

Table 5.2 Transaction Cost Comparison for eZ Cash and mCash

Transaction type	*eZ Cash (SL Rs)*	*Mobitel mCash (SL Rs)*
Cash top-up	FREE	FREE
Send money		
SL Rs 500 and below	FREE	FREE
Above SL Rs 500	5	5
Own bill payments (mobile, television, fixed line, Internet) for Dialog, Etisalat, and Hutch	FREE	FREE
Utility bill payments (LECO, NWSDB [water], etc.)	20	20
Utility bill payments (CEB)		
SL Rs 200 and below	10	10
SL Rs 200–1,000	15	15
SL Rs 1,000–5,000	20	20
Cash withdrawal		
SL Rs 200 and below	5	5
SL Rs 200–500	10	10
SL Rs 500–1,000	20	20
SL Rs 1,000–3,000	60	60
SL Rs 3,000–5,000	100	100

table continues next page

Table 5.2 Transaction Cost Comparison for eZ Cash and mCash *(continued)*

Transaction type	*eZ Cash (SL Rs)*	*Mobitel mCash (SL Rs)*
Payment of goods at appointed merchants	FREE	FREE
Mini statement	FREE	FREE
Detailed printed statement	200	200
Balance check	FREE	FREE
Change PIN	FREE	FREE
Institutional payment	FREE	FREE

Sources: "eZ Cash Transaction Charges," http://www.Ezcash.lk/pricing.php; "Transaction Limits and Charges," Mobitel mCash, http://www.mobitel.lk/mcash#Transaction Limits & Charges.
Note: CEB = Ceylon Electricity Board; LECO = Lanka Electricity Company (Private) Ltd.; NWSDB = National Water Supply and Drainage Board; PIN = personal identification number; SL Rs = Sri Lanka rupees.

eZ Cash merchants around the country, and will continue to manage this agent network, while adding any customer service networks of the new providers on request. Dialog has also added to the network the flagship stores of Etisalat and the Hutch shops that are located around the country. Dialog also manages the back office and overall customer service issues that arise during the transactions. But issues faced by customers that the two MNOs bring into the service will be the responsibility of the respective MNO.

The system is enabled under the Payment and Settlement Systems Act (No. 28 of 2005) by the Central Bank of Sri Lanka (CBSL) (figure 5.2). Hatton National Bank acts as the custodian bank for the entire eZ Cash service, and the trust agreement is administered by Deutsche Bank. A third-party service provider (KPMG)[4] acts as the customer profile manager between the eZ Cash platform and Etisalat and Hutch, so that none of the MNOs have access to the customer profiles of the other.

The cost and fee structure is the same within the interoperable system. Sending money is free below SL Rs 500 (US$4) and sending more costs SL Rs 5 (US$0.03), while the fee for receiving money is staggered and ranges from SL Rs 5 (US$0.03) to SL Rs 100 (US$0.75).[5] As can be seen by these low, affordable fees, the MNOs pass on value to consumers.

As for revenue sharing between Dialog and the other two providers, the transaction revenue from the eZ Cash service is shared with Dialog and covers the use of the eZ Cash brand, eZ Cash platform, and merchant network. Etisalat and Hutch save by not having to develop the mobile cash operation. As a result, sharing (for a fee) the eZ Cash platform is a win-win situation for all.

Developments and Possibilities

To further its national agenda on financial inclusion, in July 2015 the CBSL gave permission to Dialog to engage in the inward remittance business. Hence, for the first time, a mobile money service in Sri Lanka is able to engage in foreign remittance services. With its islandwide reach and over 16,000-plus agent locations, eZ Cash continues to revolutionize mobile payments in Sri Lanka.

Figure 5.2 Schematic of End-to-End Interoperable eZ Cash System

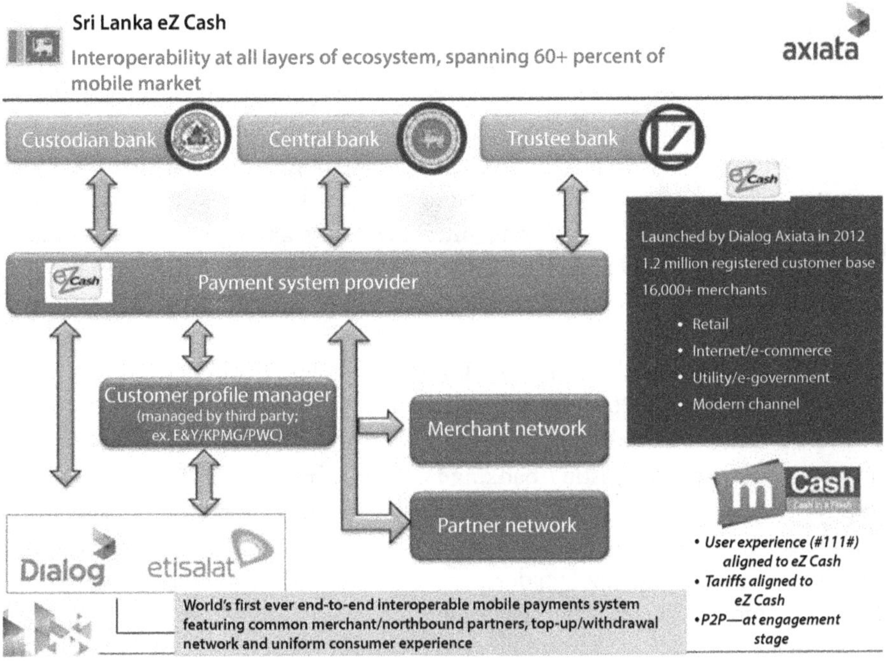

Source: GSMA 2014. © Dialog Axiata. Reproduced, with permission, from Dialog Axiata; further permission required for reuse.
Note: E&Y = Ernst & Young; P2P = person-to-person; PWC = PricewaterhouseCoopers.

Remittances continue to be important to Sri Lanka, at around 9.6 percent of gross domestic product (GDP) in 2015. Inward foreign remittances to eZ Cash mobile wallets have been made possible via the partnership between eZ Cash and HomeSend, a joint venture created by MasterCard, eServGlobal, and Belgacom International Carrier Services, which enables cross-border and cross-network value transfers. The partnership allows the migrant population overseas to transfer funds directly to eZ Cash mobile money accounts.

Sri Lanka's experience with interoperability is a success story for South Asia. In March 2015, eZ Cash beat out worldwide competition to win the Global Award for the Best Mobile Money Service at the annual Mobile World Congress in Barcelona, Spain (*DailyFT* 2015).[6] The MNO-based e-wallet system operated by Dialog enables interoperability with other telecom operators; for the custodian bank, the infrastructure is MNO-neutral. When the national switch operated by LankaClear, the national infrastructure provider, links the mobile payment systems into the main platform, the e-money–based mobile operations will be able to provide enhanced interoperable 24-7 real-time clearing and settlement facilities.

Thailand: Fully Interoperable ATMs and ADMs Easing Access

Thailand is diligently working toward systemwide interoperability. Already the country's multifunctional ATM and automated deposit machine (ADM) network is fully interoperable. As previously discussed, this has enabled nationwide proximity to cash access as well as payment transactions.

The Bank of Thailand (BOT) has established the National Payment Message Standard to support the use of corporate e-payment transactions. This standard will significantly drive interoperability and straight-through processing, as well as reduce the cost of electronic connectivity.[7] The BOT payment systems group has established a working group to develop a payment systems road map with the objective of establishing common e-money standards and developing the infrastructure for e-money to facilitate interoperability among different e-money operators. However, BOT has decided to follow a market approach in determining the feasibility of future developments.

Pakistan: Interoperability between Banks and MNOs, Even for Over-the-Counter Customers

Pakistan's Easypaisa mobile money service became the first in the country to interoperate with the existing banking structure through the Inter Bank Fund Transfer (IBFT) service, which enables customers to transfer funds between a number of banks via the 1LINK switch.[8] The IBFT service is available to both Easypaisa mobile account holders and over-the-counter (OTC) customers. Easypaisa mobile account holders can now move funds between any bank account and their Easypaisa mobile account, and OTC customers can walk into any of the Easypaisa shops and deposit cash directly into any bank account.

With 40,000 shops in more than 750 cities, Easypaisa makes access to formal financial services easier for Pakistan's nearly 6 million unique Easypaisa account holders, as well as any OTC customers. For a country that has less than 13 percent of the population in the formal financial system,[9] this is a significant step.

Tanzania: Industry-Led Interoperability

In Tanzania, use of mobile money is growing at a fast pace, as in Kenya (figure 5.3), with the regulator taking a test-and-learn approach. By enabling nonbank entities to offer payment services and working with the Tanzania Communications Regulatory Authority on the oversight framework, the Central Bank of Tanzania has shifted the regulatory approach to a mandate-and-monitor phase.

A key objective is to guide the market without stifling innovations or disrupting success, while balancing financial stability and consumer protection. Starting from a modest 1 percent adult use of mobile financial services in 2008, Tanzania experienced impressive growth to 90 percent access by 2013, with 43 percent active use among the adult population (di Castri and Gidvani 2014a).

Even more impressive is Tanzania's recent development of interoperability. Working closely with the regulator, the industry experts from payment systems and mobile money services worked through frequent meetings, debates, negotiations, and eventually consensus to successfully put together a set of standards that will govern how person-to-person (P2P) payments will be handled across networks.[10] The four main MNOs in the mobile money space (Airtel, Vodacom, Tigo, and Zantel) have come together with two of the largest banks

Figure 5.3 Comparing Mobile Money Use in Tanzania and Kenya, 2007–13

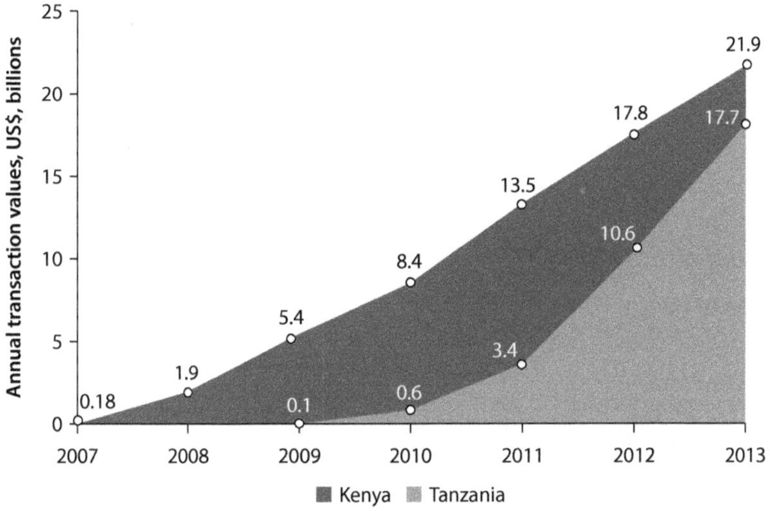

Source: di Castri and Gidvani 2014a.
Note: In 2014, US$1 = T Sh 1,627.47 (Tanzania shillings) or K Sh 86.45 (Kenya shillings).

(CRDB Bank and National Microfinance Bank) as well as the Bank of Tanzania to craft a set of operational regulations for interoperability. Once the standards are adopted, a common technical switch can be operationalized, if desired.

According to the International Finance Corporation (IFC 2015), the Tanzanian mobile market is fairly evenly distributed among the larger three players, without undue market dominance of any one player (figure 5.4). Hence, there is a healthy competitive environment and broad customer awareness of mobile payments as a tested and proven service. For the process of establishing interoperability of mobile financial services, this meant that there was relative parity in negotiating power between market actors and greater value in interoperating for both customers and providers because of the potential number of connections. Global Findex 2014 data show that the percentage of adults included had jumped from 11 percent in 2011 to 40 percent in 2014, with 32 percent included through mobile accounts.[11] If interoperability works as planned, Tanzanians will not only be able to send money through the network but also connect to banks as well.

In the meantime, Tigo and Vodacom signed the mobile money interoperability agreement in February 2015, thereby connecting M-Pesa and Tigo Pesa customers in Tanzania. Nearly 4 million Tigo Pesa customers and 7 million M-Pesa customers will be able to transact with each other. The process is still ongoing, and it remains to be seen how much fruit this collaboration will bear. Another important interconnect collaboration is happening between the East Africa region's two biggest mobile money operators, Vodafone Group and MTN Group. This will enable convenient and affordable international remittances between M-Pesa customers in the Democratic Republic of Congo, Kenya, Mozambique, and Tanzania and MTN Mobile Money customers in Rwanda, Uganda, and Zambia. Hence, interoperability seems to be picking up in the Africa region.

Figure 5.4 Active Subscriber Market Shares of Tanzanian Mobile Service Providers, 2014

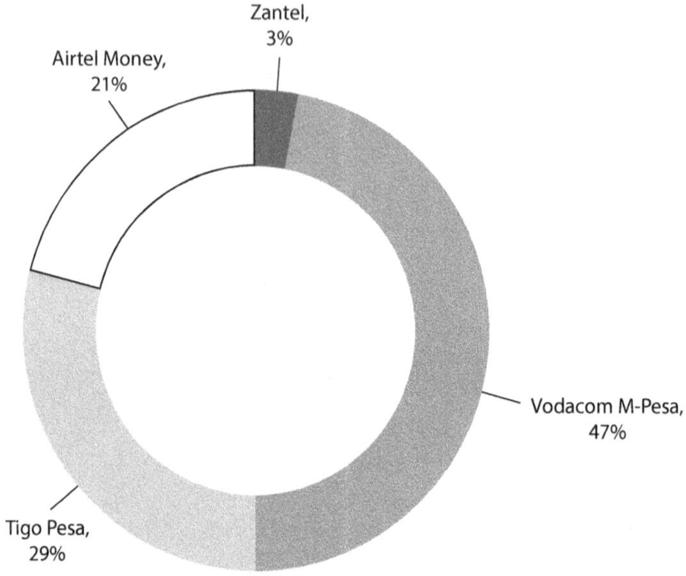

Source: IFC 2015.

Indonesia: Multi-Wallet Interoperability

Indonesia is the world's fourth most populous country, with a population of around 245 million, over 60 percent of whom are of working age. Indonesia has the same archipelagic barriers to connectivity and access that Maldives and the Philippines have, but on a much larger scale. Indonesia has over 17,500 islands, of which over 6,000 are inhabited and around 1,000 are permanently settled. Driven largely by domestic demand, Indonesia's economic growth is on an upward trajectory, and connectivity and financial access are becoming increasingly important to Indonesians.

The Global Findex 2014 data show that only 36 of adult Indonesians have access to the formal financial sector;[12] of those, 27 percent save at a financial institution, while 13 percent borrow. Given the country's dispersed geographic structure, connectivity to formal financial institutions via traditional bank branch and ATM networks will always remain a problem. Although Indonesian mobile penetration is comparable to its peers in the region, its banking services are relatively underdeveloped (figure 5.5). With mobile-phone penetration at over 100 percent, Indonesia is in the perfect position to use mobile money to enhance financial inclusion. Indonesia currently has six mobile money schemes.[13]

Regulatory Bottlenecks

Regulatory barriers have inhibited the growth of agent and branchless banking in Indonesia. The central bank, Bank of Indonesia (BI), did not allow agency banking. Any person providing money transfer or cash withdrawal service needed

Figure 5.5 Financial Account and Mobile-Phone Penetration, Indonesia versus Selected Asian Countries, 2014

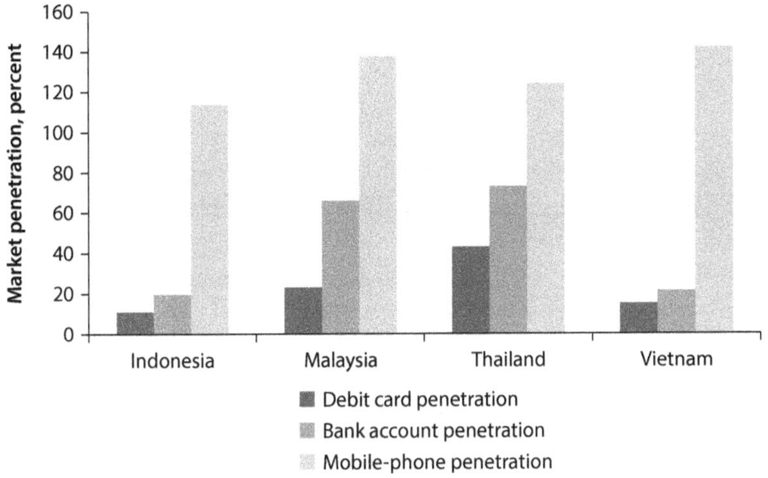

Source: Ernst & Young 2014.

a money transfer license. Also, outsourcing of opening accounts or know-your-customer (KYC) assessments was not allowed, and no concessions were made for the inherently low risk of low-value transactions.

Although regulations have been issued allowing nonbanks to offer e-money, until recently, any person wishing to start a money remittance or cash-out service had to obtain a license as a money remitter. This was not feasible for small businesses, which were discouraged from applying for licenses. As such, people in the outer islands and rural communities had severely limited opportunities to obtain financial services. To withdraw money from an e-wallet, a person has to go to an outlet managed by their own operator; with only around 25 outlets nationwide for each operator, mobile money was not a viable option in rural areas. Realizing the regulatory bottleneck, however, BI issued Fund Transfer Regulations in 2013, effectively paving the way for mobile money agents to offer cash-in/cash-out and account-opening services. As a result, thousands of agents started signing up (Camner 2013).

Industry Collaboration on Multi-Wallet Interoperability

Realizing that they have more to gain by working together and pooling resources than by competing with each other to cover the geographically challenging terrain, discussions among three MNOs—Telkomsel, Indosat, and XL (Axiata)—led to the decision to interoperate. Recognizing the opportunity present in the payments market, these MNOs decided to interoperate yet keep their own identities by way of a multi-wallet interoperable arrangement.

On May 15, 2013, just six months after the discussions started, the groundbreaking scheme was launched, allowing customers of the three MNOs to transfer and cash out money from any location across each other's network. Cash-ins are currently a cost for each mobile money deployment but may lead to other more lucrative businesses, such as bill paying, other payment services, banking, and so on.

Most important in this operation is the continuous dialogue among the members. They still need to iron out many issues, some of them industrywide, others system-specific. A primary concern is the protocol for customer grievance and redress. They have already decided to address each operation bilaterally, and end-of-the-day net settlement is to be done by the custodian bank in real time with the BI payment system. Managing Anti-Money Laundering and Combatting Funding of Terrorism (AML/CFT) regulations and training staff to perform KYC and customer due diligence (CDD) duties in a tiered fashion and moving toward risk-based supervision are other important aspects.

These measures are expected to lead customers to develop faith in the system and, over time, to transition toward e-money transactions. Anecdotal evidence suggests that many people still do not understand the value of mobile banking. A survey carried out by Financial Inclusion Insights (FII 2015) found that only around 3 percent of Indonesians were aware of the concept of mobile money, and those who were aware had heard of only one type of service (figure 5.6). Awareness raising is very important.

Figure 5.6 Mobile Money Awareness in Indonesia, 2014

Indonesian Inclusion

Many Indonesians don't have access to banks, but they do have access to mobile phones that could provide financial services—yet few have ever heard of such mobile services.

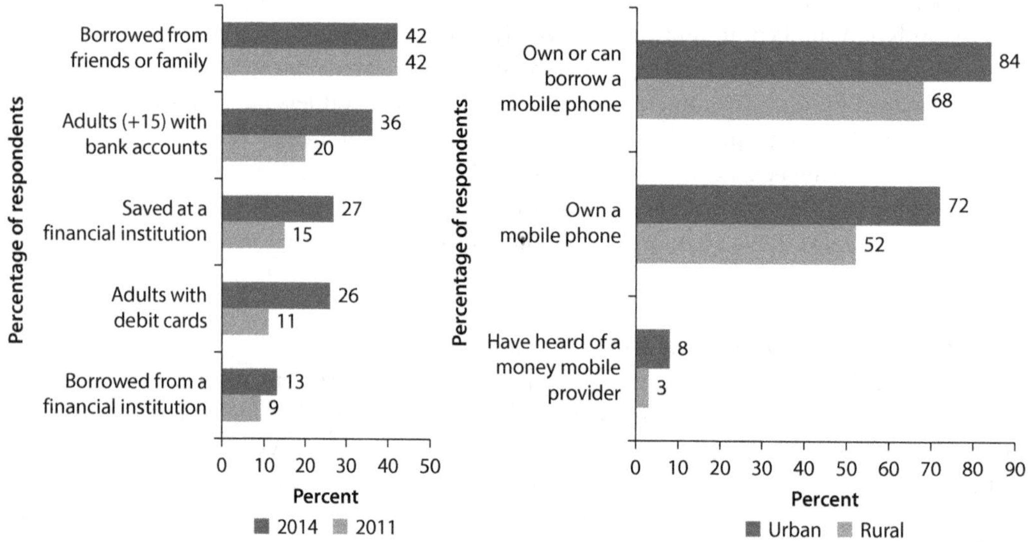

Source: Schonhart 2015. Adapted with permission from Dow Jones, Inc. Further permission required for reuse.
Note: Left chart: Survey conducted of 1,000 respondents May 18–13, 2011, and May 3–June 4, 2014. Margin of error is +/−3.6 percentage points. Right chart: Survey of 6,000 respondents conducted August–November 2014. No margin of error provided.

Banks in Indonesia are fully interoperable and have established their payment and clearing services and ATMs. However, agent banking is yet to pick up. The majority of ATM transactions are actually payments, not cash withdrawals, indicating that there is a demand for self-service electronic payments. Opening up interoperable services for MNO-led e-money is a smart early move, and MNOs have gained on the banks. By building consumer trust and awareness, they can secure this potentially untapped market. Either way, with a broader range of competitive services to choose from, consumers will be the winners.

Agent Network Management in Kenya

> *You can buy a Coca-Cola anywhere in the world, but affordable products that provide essential value like water treatment or lighting often do not reach billions of poor populations around the globe. However, in what is commonly known as the "last mile distribution challenge," some social entrepreneurs are providing innovative solutions to make the last mile a first opportunity.*
>
> —Nicolas Chevrollier and Stéphanie Schmidt, "Overcoming the Last Mile Challenge: Distributing Value to Billions"

The world's best innovative solution or product could fail to achieve its objectives if the last mile problem is not solved in a cost-effective manner. Making the final connection between consumers and the service or the product has often proven to be disproportionately expensive to solve, or sometimes may be forgotten in the focus on technological and other important problems. Chevrollier and Schmidt (2014) sum up this reality:

> A majority of the population in developing economies live in rural areas often accessible only by poor quality road infrastructure. Furthermore, geographical isolation or limited access to relevant information disconnects populations in many developing countries from any business value chain. The consequence—which can affect both urban and rural populations—is that products providing essential value either do not reach the intended customers or are more expensive or lower quality than the standard products that are accessible by other populations.

These challenges hold for financial services as well. Comparing case study experiences of success in e-money deployments in some cases with inability to reach a seemingly feasible target in others, the manner in which the last mile problem is addressed is a critical element that often makes the difference. Even if all the key elements appear to be in place—progressive regulators pass enabling regulations, an innovative technology model exists that would be seamless in operation, and there is supporting physical infrastructure or connectivity—if the human element of the customer is not taken care of, the intended goal of widespread, low-cost access to financial services by the poor may not be achieved.

According to the GSMA (2014), on average, there were 2.3 million mobile money agent outlets globally in developing countries in 2014 (figure 5.7). This is 4.3 times the average density of bank branches in these markets, which

Figure 5.7 Number of Financial Access Points across Developing Countries, 2014

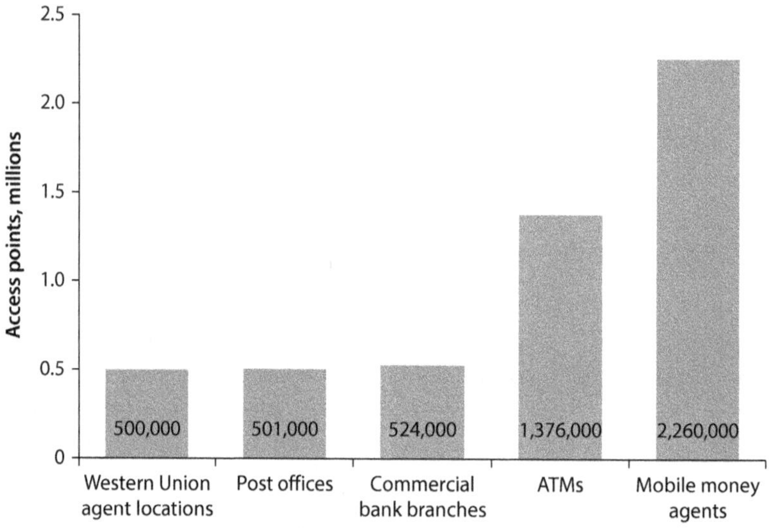

Source: GSMA 2014.
Note: ATMs = automated teller machines. GSMA data from the Mobile Money for the Unbanked (MMU) Deployment Tracker, the 2014 Global Adoption Survey of Mobile Financial Services, and MMU estimates and forecasts. The countries surveyed included those classified by the World Bank in 2014 as "developing countries" plus four countries that were not on that list: Chile, Qatar, Singapore, and the United Arab Emirates.

total 524,000. In three-quarters of the 89 markets where mobile money is available today, agent outlets outnumber bank branches. In 25 of those markets, there are more than 10 times as many mobile money outlets as bank branches.

The ability of the MNOs to reach the customer and their last mile advantage is undeniable, as they already have extensive touchpoints by way of airtime agents. Although that might bode well for the success of any MNO-led mobile money initiative, this is not necessarily the case. While the number of registered accounts highlights the growing ubiquity of mobile money, the number of active accounts is more important to understand the speed at which customers are adopting mobile money services. At the end of 2014, out of 255 live mobile money initiatives across 89 countries, 21 services had more than 1 million 90-day active accounts, seven of which passed this threshold during 2014 (GSMA 2014).

For any mobile money or digital financial service to reach the base of the pyramid, therefore, building efficiency and quality into the agent network is a critical element in success. The agents perform registration, facilitate cash-in/cash-out, and also act as brand managers and awareness builders. Hence it is important that they are recruited wisely and trained on how to perform KYC and CDD duties, adhere to AML/CFT standards, manage liquidity, address consumer complaints, build awareness, and promote the service. Interpersonal skills are important, too: one main advantage of these last mile touchpoint agents often is that they are already well known in the area that they serve.

As mobile money initiatives grow in scale, the challenge of managing the network of agents is compounded, possibly stalling growth if not carefully structured. Examining M-Pesa's agent network management offers crucial insights into how Safaricom (the mobile network operator that developed M-Pesa) recruits, trains, and manages its vast network of agents to reach high levels of efficiency.

Winning with Superior Agent Network Management

The dominance of M-Pesa in Kenya is partly due to the efficient management of its vast network of agents. Today M-Pesa has 26 million users (around 18 million active, as against 35 million deposit accounts in the entire banking sector), who transact through nearly 130,000 agents, with a deposit base of K Sh 1.8 billion and K Sh 7 million in loans issued monthly (CBK 2016; Maina 2017). The fact that M-Pesa has raised Kenya's financial inclusion from 26 percent to 70 percent underscores the importance of the whole agent operation (Johnson 2012).

Agent Recruitment

Before the launch of M-Pesa, it was difficult to imagine the entrance of an MNO into the field of financial services. The existing network of dealers that Safaricom used to distribute airtime seemed the obvious choice to facilitate M-Pesa transactions by providing cash-in/cash-out (CI/CO) services. An airtime dealer is an independent company, typically with 5–20 retail outlets, selling airtime, mobile phones, and other goods. These dealers quickly agreed to become M-Pesa agents because of their trust in the Safaricom brand. Agents have to register the customers by performing KYC/CDD checks, educate customers on available digital options, and perform CI/CO functions.

To support rapid growth, Safaricom initially had to ensure that enough agents were recruited to provide a viable service for users and, subsequently, balance the number of agents with the numbers of customers to ensure that agents had profitable businesses while customers did not face overcrowding and long waiting times. The service was going to be successful only if a large network of agents and customers was quickly established. Despite some initial lag in the growth of agents relative to the growth of customers, engagement of agents picked up in 2009 and brought the agent-customer ratio down from a peak of 1 to 1,000 in mid-2008 to about 1 to 600 by the end of 2009, and subsequently kept pace (figure 5.8) (Davidson and Leishman 2012; Jack and Suri 2014). If Safaricom had gotten the pace of scale-up wrong, M-Pesa could easily have failed.

In the early stages, Safaricom was able to use its reputation to bridge the gap in trust that is typical for new products and services. Agents trusted that Safaricom could launch a new profitable service, and consumers trusted Safaricom's brand enough to experiment with the new system. Safaricom enjoys greater consumer confidence and trust than do many banks, and it is still Kenya's most admired brand. To support the launch and growth of M-Pesa, Safaricom needed to build an ecosystem of actors. Each actor needed to be incentivized for Safaricom to grow M-Pesa without compromising the quality of customer service.

Figure 5.8 Growth in Number of M-Pesa Customers and Agents, 2007–14

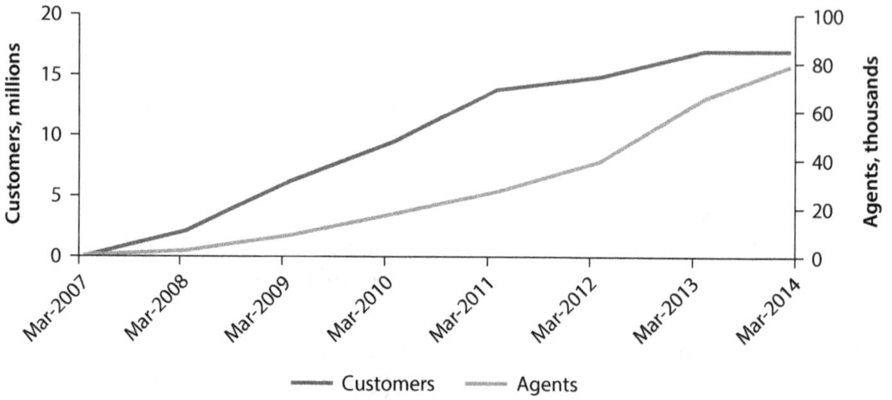

Source: Safaricom.

Table 5.3 Roles and Responsibilities in M-Pesa Agent Structure

Roles and responsibilities	*Safaricom/ M-Pesa*	*Superagent*	*Aggregator*	*Agent*	*Agent network managers*
M-Pesa business strategy	√				
Selecting agents	*		√		*
Providing equipment				√	
Providing start-up capital				√	
Agent rebalancing		√	*		
Agent training			*		√
Agent monitoring			*		√

Source: World Bank.
Note: √ = primary responsibility; * = some involvement.

The roles and responsibilities of these different actors have necessarily evolved over the course of M-Pesa's existence. That evolution has been pivotal to M-Pesa's success as it grew from start-up (albeit with sound corporate backing) into a major provider. The roles and responsibilities of agents, aggregators, agent network managers (ANMs), and superagents are summarized in table 5.3 and detailed in table 5.4.

Incentive Structure

Since M-Pesa's launch, agents have been incentivized to acquire customers for M-Pesa with a K Sh 40 (US$0.47) commission paid when they sign a customer and a further K Sh 40 when that customer makes his or her first deposit into an M-Pesa account. A staggered commission is also given to agents for each deposit that they receive and each withdrawal that they facilitate. It is free for customers to sign up and free to deposit, with charges levied on withdrawals and P2P payments.

Safaricom invested heavily in its agent channel. Commissions to agents to register 4 million customers cost the company US$5 million (Mas and Morawczynski 2009).

Table 5.4 Detailed Roles and Responsibilities in M-Pesa Agent Structure

Safaricom M-Pesa
Issues e-money 1:1 against cash held in M-Pesa trust accounts

Superagent
- Direct relationship with Safaricom (usually a bank)
- Purchases e-money from Safaricom and sells to agents
- Accepts deposits of cash from agents
- Retains 1% commission on e-float sales
- Deposits funds into M-Pesa trust account

Aggregator
- Agent with direct relationship with Safaricom
- Purchases e-money from superagent (or Safaricom)
- Deposits cash with superagent (or into M-Pesa trust account)
- Recruits, trains, and monitors a group of agents (subagents)
- Manages cash and e-money for group of agents
- Retains 20% of subagent's commission

Agent
- Agent with direct relationship with Safaricom
- Purchases e-money from superagent (or Safaricom)
- Deposits money with superagent (or into M-Pesa trust account)
- Registration of M-Pesa customers
- Depositing cash into registered customers' M-Pesa accounts
- Processing cash withdrawals for registered M-Pesa customers
- Processing cash withdrawals for nonregistered M-Pesa customers
- Customer education
- Compliance with Safaricom AML and KYC policies
- Compliance with Safaricom business practices
- Branding of their outlets per Safaricom-provided guidelines

Subagent
- Registered with Safaricom
- Purchases e-money from its aggregator
- Deposits cash with aggregator or into bank account of aggregator
- Customer care and compliance as per agent

Agent network managers
- Firms tasked by Safaricom to provide agent training and monitoring
- Check branding of M-Pesa agents
- Circulate branding materials, posters, and M-Pesa registers
- Carry out mystery shopper checks on AML and KYC compliance

Source: World Bank.
Note: AML = anti-money laundering; KYC = know your customer.

Even agent aggregators are incentivized to support customer growth with a K Sh 10 (US$0.12) commission for each new customer their agents sign (Omwansa and Sullivan 2012). Although the M-Pesa commission offered to agents is lower than that for airtime sales, M-Pesa still makes sense for agents.

The business case for agents has been important since the launch of the service. A Consultative Group to Assist the Poor (CGAP) study of rural and urban M-Pesa agents 18 months after the launch of the service revealed some of the reasons to become an agent (Mas and Morawczynski 2009). The noncommission benefits include being associated with Kenya's best-known brand and attracting increased foot traffic in shops. This was and remains particularly important for rural agents, where M-Pesa brings cash into the villages; the cash is spent locally, usually in an agent's shop.

M-Pesa generated nearly three times more revenue than mobile-phone airtime revenue for the typical agent. Average daily commissions for selling airtime were US$3.78 compared with average daily commissions of US$10.65 from M-Pesa transactions. (This had to be offset by the agent's cost of travel to the nearest bank branch, which cost up to US$5, as well as lost revenue if the shop was closed during travel.)[14]

The business case for individual agent outlets depended on their costs of rent and wages (if it was not an owner-operated business), the return on their capital tied up in the M-Pesa float, and any losses to theft. At the average number of transactions, an M-Pesa agent needs US$1,341 to maintain his or her float and cash balances. The cost of capital may be a borrowing cost or an opportunity cost. Many aggregators strongly discourage borrowing to maintain working capital, because the additional expense undermines the profitability of the business. For the average agent, the largest proportion of costs is associated with rebalancing—that is, management of liquidity between cash and e-float—estimated at US$1.13 per daily transaction; although the introduction of more superagents may have lowered this cost over time. The average M-Pesa agent was profitable at 53 transactions per day, but only if wages and rent were low.

Awareness Building

At the same time that it was incentivizing agents to promote the service, Safaricom invested in a massive advertising campaign with the simple proposition: "Send Money Home." The advertising budget was spent in a big splash to support the rapid uptake of service. M-Pesa offered just three features: the ability to cash-in and cash-out at agents' locations; send money P2P; and buy airtime direct from Safaricom.

As M-Pesa began to take off, Safaricom had to invest in regular system upgrades. This massive investment can be seen in Safaricom's financial results for fiscal year 2008, the first year of M-Pesa's launch: capital expenditures were up by 41 percent, and sales and advertising were up by 92 percent. To execute this massive project, Safaricom had ensured buy-in from senior management, including from the chief executive officer (CEO), and established a dedicated business unit to manage M-Pesa. It is estimated that Safaricom spent US$30 million over the first three years to launch M-Pesa (Omwansa and Sullivan 2012).

Agent Structure

MNOs such as Safaricom have a built-in advantage as mobile money service providers because of their experience in building and managing agent networks for the distribution of airtime. Safaricom's approach to agent network management grew from airtime distribution, but it evolved over time to ensure a consistent customer experience and a viable business proposition for agents. As M-Pesa grew, Safaricom needed to constantly balance the number of agents with the number of customers; this was to ensure high enough commissions for agents without overcrowding outlets so much as to cause customer dissatisfaction.

Figure 5.9 Initial M-Pesa Agent Network Structure

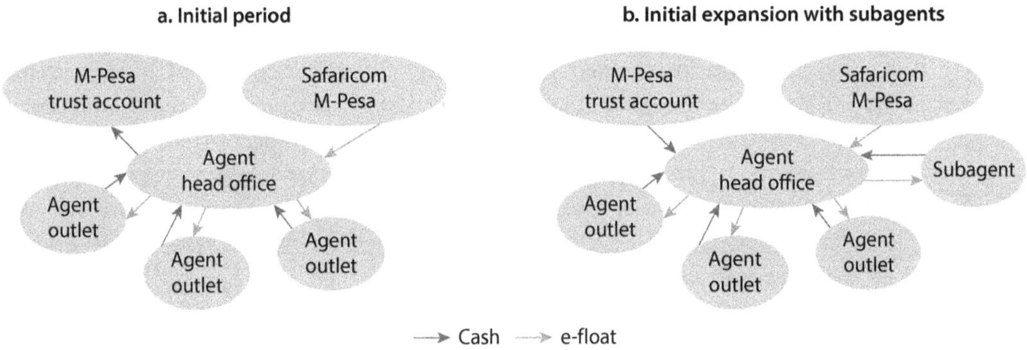

Note: Blue arrows indicate deposit of cash. Orange arrows indicate e-float crediting.

The initial arrangement of players in the M-Pesa system is shown in figure 5.9, panel a. During this initial period Safaricom directly recruited M-Pesa agents from its pool of existing airtime dealers. These agents offered M-Pesa service through the outlets that they directly owned. In these early days, M-Pesa agents were required to have an outlet in at least three provinces; this was later relaxed to three outlets, and they could have subagents (figure 5.9, panel b). Safaricom maintained a direct contractual relationship with all agents at this time. Figure 5.9 also shows the flows of cash and e-float between the various parties, a process described in the "Expanding E-float and Cash Management" subsection.

Outsourcing Agent Network Training and Management

In the first couple of years after launch, there was exponential growth of both agents and customers. Safaricom centrally controlled the agent network in the very early stages of M-Pesa, but the rapid growth in agents quickly led it to outsource critical components of agent management. M-Pesa had 10 regional managers across the country who were responsible for agent management. But with the massive expansion of the network, regional managers now primarily manage firms tasked with agent management.

These contracted ANMs helped Safaricom scale their agent network quickly, while continuing to provide high-quality, consistent service to customers. The first ANM that Safaricom recruited was a Kenyan marketing firm, Top Image, which was given the specific responsibility of training agents, distributing marketing and business material, and monitoring agents. Top Image also helped to design a training curriculum.

Training standards are defined by Safaricom and implemented by the ANMs. The ANM trains new M-Pesa agent staff on how to operate M-Pesa successfully and explains how to manage liquidity and how to comply with KYC and AML rules and requirements. Then a simple test is used by the ANM to determine whether prospective agents are ready to serve customers, in terms of both staff and outlet readiness. Top Image is responsible for the delivery of marketing and business materials to all M-Pesa outlets. Business tools include the agent tills, transaction books, and registration books. They also distribute marketing and branding materials and make sure that these are properly displayed at all M-Pesa outlets, including M-Pesa tariff posters.

By 2011 Top Image was monitoring more than 80 percent of M-Pesa agents (23,000). At the time, Top Image had a team of 92 trade development representatives across the country monitoring M-Pesa agents (250 agents per trade development adviser). During fortnightly visits these representatives use a 10-point checklist to assess the agent outlet. Items that are checked include whether the outlet has enough cash and e-float and whether the logbooks are in good order. They also carry out "mystery shopper" checks on their KYC and AML procedures.

Safaricom now uses a number of ANMs in addition to Top Image to provide it with information about the M-Pesa agent network and its performance. Firms like Top Image are paid a fixed fee for their service, rather than being offered a commission split. This is to ensure that they independently monitor agent outlets, and that there is not a disincentive to report poor compliance—with Safaricom's KYC and AML procedures, for example.

As businesspeople clamored to become M-Pesa agents, existing agents made informal arrangements with outlets that they did not own or operate (figure 5.10). While these subagent outlets were not owned by the original M-Pesa agent, they served to supplement the outlet distribution. Thus the roles of aggregator and subagent evolved organically as an informal arrangement that had the tacit approval of Safaricom. Among the more than 10,000 M-Pesa agents, over 50 percent were subagents of aggregators. Agents acting as aggregators usually retain 20–30 percent of subagents' commissions, but sometimes as much as 50 percent. There were a number of problems with subagents, who did not have a direct relationship with Safaricom, in particular inconsistent branding, inadequacy of float, and poor

Figure 5.10 M-Pesa Agent Network Structure with Formal Introduction of Aggregators

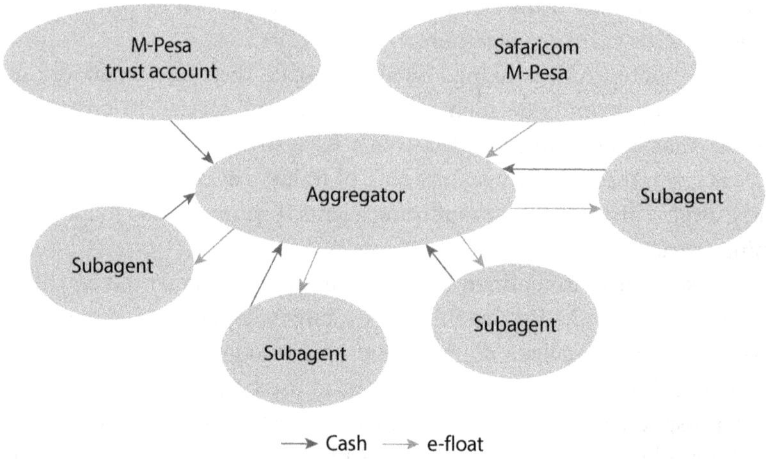

Note: Blue arrows indicate deposit of cash. Orange arrows indicate e-float crediting.

knowledge of AML procedures. These challenges resulted in a further refinement of Safaricom's agent network structure.

In late 2009 Safaricom began to reassert its control over the entire agent network (Flaming, McKay, and Pickens 2011). The role of aggregator was officially recognized. Aggregator agents were tasked with an active role in subagent training and monitoring. At the same time, the maximum portion of commission that an aggregator could take from their subagents was capped at 20 percent. In addition, subagents were required to establish a direct contractual relationship with Safaricom. Given some of the complaints that subagents had about unscrupulous aggregators, Safaricom also mandated that a subagent could become an agent in their own right after three months with their aggregator.

Aggregators identify potential subagents according to their own criteria, such as outlet location, whether there is a large customer base, and whether the area is undersupplied by existing M-Pesa outlets. ANMs like Top Image then vet applicants and provide initial training. The incentive for aggregators is to recruit and manage agents who generate profits from high volumes of business. The economics differ between an aggregator's own locations and those of a subagent. An aggregator location needs sufficient revenue to cover the cost of at least one full-time employee and rent, these costs being higher in urban areas than rural areas.

A CGAP study looked at the business case in 2011 when the average number of transactions per day for an agent outlet was 53 (Flaming, McKay, and Pickens 2011). An aggregator's own rural agent could generate profits of US$60 on 53 transactions, but an urban aggregator agent could not generate profits until 80–100 transactions had been carried out. The 20 percent commission share on subagents represents US$41 profit for an aggregator from subagents with 53 transactions per day. The average aggregator with 100 agents is quite profitable, but this profitability rests on owning a large proportion of the agent outlets that they manage.

CGAP's financial modeling shows that customer transaction fees provide the primary source of revenue within M-Pesa. M-Pesa retains 58 percent, aggregators take 8 percent, and agents receive 34 percent. In addition, Safaricom profits from airtime sales on the M-Pesa platform, at the same time making savings on airtime distribution. There are additional benefits to Safaricom from increased average revenue per user and reduced churn rates.

Expanding E-float and Cash Management

Agent recruitment is vital to M-Pesa, and so is the efficient running of their daily business: the management of their stocks of cash versus e-money (a relationship called "e-float"). Safaricom required K Sh 100,000 in working capital from prospective agents, and this was split 50-50 between e-float and cash. The typical urban-to-rural remittance pattern in Kenya means that an agent in an urban location generally has more deposits and hence a greater need for e-float, while a rural agent typically handles a larger number of withdrawals and needs greater stocks of cash. If an agent runs out of e-float, he or she will not be able to accept any more cash deposits, and an agent who has run out of cash cannot process cash-outs. The management of liquidity is essential to ensure continuous customer service. Hence, Safaricom provides training on this essential activity. Agents manage liquidity by depositing cash in a bank account in exchange for e-float. This active management of liquidity is referred to as "rebalancing."

Safaricom issues e-money to agents on a 1-to-1 ratio against the money held in the M-Pesa trust account; hence, no money is created. Since M-Pesa's inception, the funds of all M-Pesa customers have been held separate from Safaricom in a trust account held by a prudentially regulated bank. The Commercial Bank of Africa (CBA) held the first M-Pesa trust account. (CBA continues to hold an M-Pesa trust account, but the monies are now also split between three additional banks in Kenya to reduce risk.) In the early days, Safaricom was centrally issuing e-float to aggregators, who then distributed it to their subagents. The process did not appear to be well automated, as delays of two to four days in issuing e-float were regularly reported by agents. This tied up agents' working capital and limited the volume of business that they could support.

To provide greater liquidity and much speedier rebalancing for agents, Safaricom introduced the "superagent" concept (figure 5.11). In May 2009, Safaricom unveiled its first M-Pesa superagent, Kenya Commercial Bank (KCB). Safaricom also made an agreement with KCB to provide collateral backing for agents at favorable rates. KCB additionally provided overdraft protection

Figure 5.11 Current M-Pesa Agent Network Structure and E-float/Cash Management Process

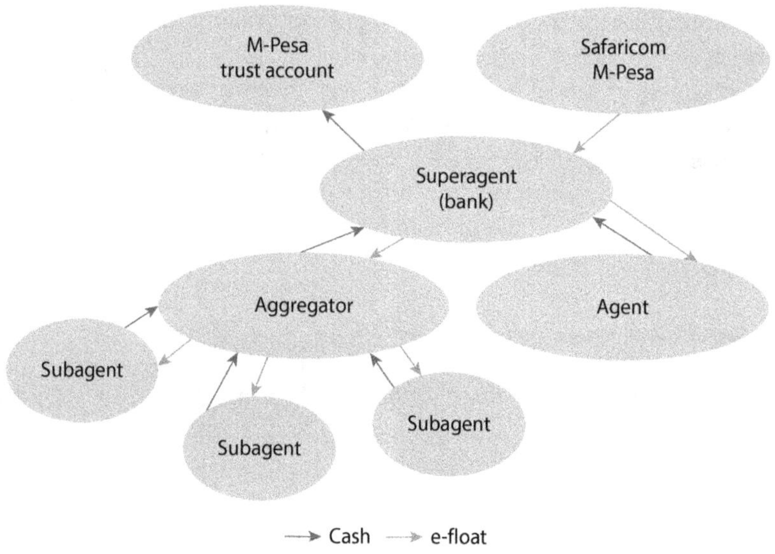

Note: Blue arrows indicate deposit of cash. Orange arrows indicate e-float crediting.

for agents' working capital delivered through M-Pesa. Safaricom has continued to recruit banks as superagents to provide vital liquidity services to the agent network. Superagents deposit funds directly into an M-Pesa trust account and are issued with e-float by Safaricom, which they then sell to agents for a 1 percent commission. Although it is still possible for agents to buy e-float directly from Safaricom, in practice this is done infrequently. Banks, acting as superagents, provide agents with dedicated tellers in their branches and instant delivery of e-float after a cash deposit. Barclays Bank and Ecobank joined as superagents shortly after KCB, helping to formalize a new financial ecosystem of which banks are an essential part. (Superagents were then also able to offer bulk payment services to their corporate customers via M-Pesa.) Most agents rebalance their cash/e-float holdings daily with the help of the bank network.

Another tool to assist agents in e-float management is the management SIM, which enables aggregators to speedily credit e-float to subagents. Aggregators can use an online interface that allows for more streamlined and speedy e-float management. Subagents either deposit float with their aggregator or directly into the aggregator's bank account. Through the management SIM, the aggregator is notified when a subagent has made a cash deposit at the bank, and hence they immediately release e-float. This eliminates the delay incurred when waiting for subagents to present teller deposit slips to the aggregator. The speed of rebalancing allows agents to be more flexible with their working capital; if they are in an urban area, where the predominant activity is taking deposits, they can hold more in e-float.

To summarize, Safaricom controls the business strategy for M-Pesa and sets the rules for the various actors. Safaricom defines the requirements for becoming an agent and manages the overall recruitment process, but it is the aggregators who select and manage these agents for vetting by the ANM and for approval by Safaricom. Agent training and monitoring is outsourced to an ANM like Top Image. The agents are responsible for providing their own phone and start-up capital. Banks act as superagents to rebalance agent e-float and cash levels on demand.

Because each market is different, it is hard to generalize the M-Pesa model and expect that a similar mobile deployment would necessarily thrive in another country environment. What is more important is to appreciate how M-Pesa understood the agent network as a critical enabler and evolved it dynamically to suit the market's needs. The levels in the network were not predetermined but evolved to fill a gap or to formalize a practice that seemed to work. Safaricom was a dynamic, hands-on manager that realized early on that, left to its own devices, sooner or later the market would find a solution, and ensured that this took place within established boundaries.

Mobile Money Applications as the Digital Ecosystem

Many emerging markets lack the core infrastructure (national switches and settlement systems, connectivity, even electricity) required for innovative payment systems to operate reliably. One option that addresses this pain-point is Eseye's innovative machine-to-machine (M2M) technology. By embedding SIM chips into solar-powered lighting units, Eseye has created a remote device management system enabling prepaid and pay-as-you-go (PAYG) financing via M-Pesa in East Africa. Prepaid customers buy exactly as much light as they need, while PAYG customers see their lights go out if they miss a payment. The payment options enabled by Eseye's technology and its relationship with M-Pesa circumvent infrastructure short-comings in East Africa by operating through mobile channels. Without this technology, customers would have to pay up front or borrow in order to purchase the system. With it, many more people are gaining access to solar lighting.

—Monica Brand Engel and Jackson Scher, "Four Barriers—and Four Solutions—to Financial Inclusion through Payment Innovations"

Competition is increasing in many markets as mobile money becomes a main-stream product for a growing number of operators. Two or more mobile money services operate in 56 markets, while 36 markets have three or more services (map 5.1). In three markets—Pakistan, Sri Lanka, and Tanzania—MNOs interconnected their services following the example of MNOs in Indonesia, where domestic interoperability was implemented in 2013 (GSMA 2014). Hence the question is no longer whether mobile money services are available, but how to ensure that the industry continues to grow sustainably.

Map 5.1 Number of Live Mobile Money Services for the Unbanked, by Country, 2014

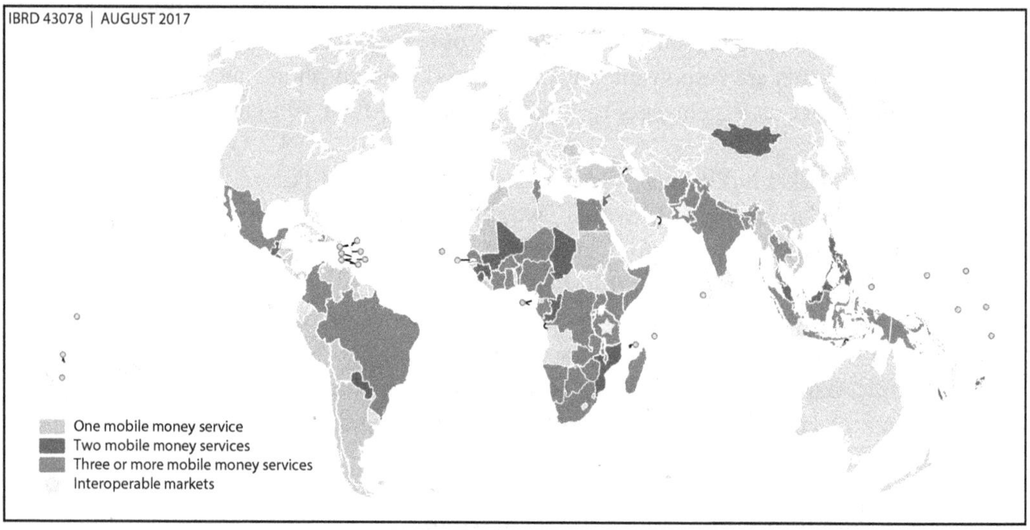

Source: GSMA 2014. © GSMA. Reproduced, with permission from GSMA; further permission required for reuse.
Note: GSMA data from the Mobile Money for the Unbanked (MMU) Deployment Tracker, the 2014 Global Adoption Survey of Mobile Financial Services, and MMU estimates and forecasts. The countries surveyed included those classified by the World Bank in 2014 as "developing countries" plus four countries that were not on that list: Chile, Qatar, Singapore, and the United Arab Emirates.

Figure 5.12 Mobile Money Transfer Value Chain

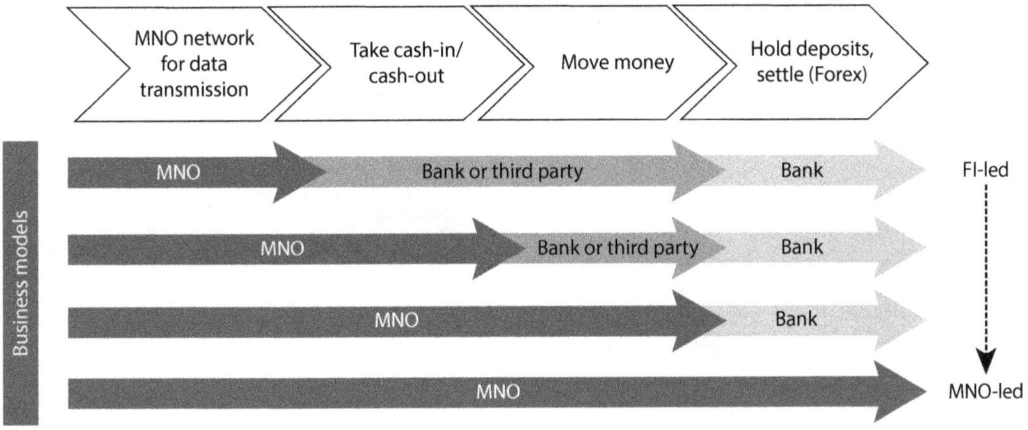

Source: Gencer 2009. © mPay Connect. Reproduced, with permission from mPay Connect; further permission required for reuse.
Note: FI = financial institution; MNO = mobile network operator.

Sustainability of mobile money services need not focus solely on providing the basic functions of CI/CO, bill payment, and money transfer. Moving along the value chain for an MNO starting from simple CI/CO to becoming a full-fledged banking agent is a possibility, depending upon the regulatory framework in a country (figure 5.12). The possibility of being a platform for the digital ecosystem for millions of subscribers is an enticing prospect. Government payments, social grant programs, and payrolls are some of the services that could be delivered via mobile money platforms. In Kenya, applications (apps) have been designed by independent private sector developers. In partnership and by fee-sharing agreements with Safaricom, M-Pesa has provided the infrastructural foundation upon which these value-added services have been built, thereby greatly extending M-Pesa's digital connections and thus Kenya's movement toward a cash-lite society.

Innovative Mobile Applications Enabling Movement toward a Cash-Lite Economy

Mobile payment–based e-money solutions can be operated even under trying geographical circumstances and locations. Unlike ATMs, point-of-sale (POS) operations, and Internet and telephone banking services, mobile e-wallet solutions do not require broadband connectivity. Most modern-day mobile money solutions are operated using simple 2G phones—usually available at a cost of less than US$20 dollars per handset and thereby affordable to poor people. Because of the simplicity of the technology—requiring neither Internet service nor 3G or 4G connectivity—it is possible for MNOs to provide e-money services in faraway and outer islands and atolls in countries like Maldives and the Philippines.

Figure 5.13 M-Pesa Service Development, 2007–13

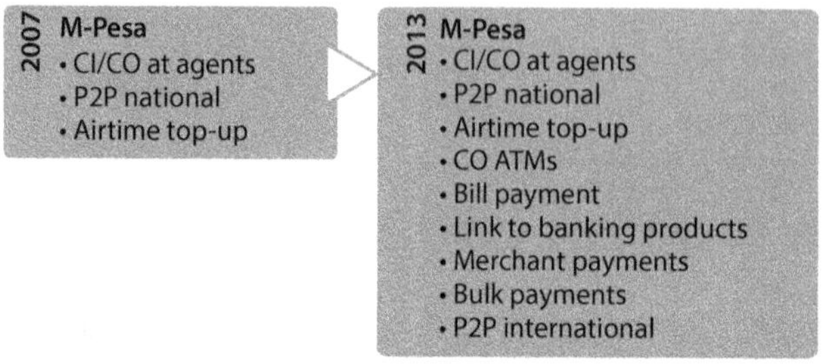

Source: World Bank and Safaricom 2017 data.
Note: ATMs = automated teller machines; CI/CO = cash-in/cash-out; CO ATMs = cash-out-only ATMs; P2P = person-to-person.

M-Pesa was launched with an intentionally basic functionality, focused on CI/CO, P2P, and buying airtime. This provided a clear proposition that agents and customers could understand and adopt. Today, as figure 5.13 details, the functionality is significantly richer. As functionality was added, it opened opportunities for other businesses, particularly financial institutions, to use the system.

Table 5.5 shows the timeline of M-Pesa expansion of functionality and services, and some of the most important enabling applications are summarized in subsequent pages. Since M-Pesa's launch in 2007 Safaricom has continued to enrich the offer with additional services and features. This customer-centric approach has increased the utility of M-Pesa for more and more users. The key service enrichments (table 5.5) highlight those services that were introduced onto the platform following negotiations and agreement by private developers with Safaricom (some were not successful).

It is also useful to note that Safaricom has a financial incentive for the expansion of the digital ecosystem that also well serves the customer, micro- and small businesses, and microfinance institutions (MFIs) as well as the economy. In this regard, commissions to M-Pesa agents are now the largest direct cost of running M-Pesa. Safaricom's 2013 first-half results show that K Sh 5 billion (US$60 million) was paid out to M-Pesa agents, compared with voice and short message service (SMS) interconnect costs of K Sh 3.09 billion and airtime commissions of K Sh 4.65 billion.

Given the high cost of incentivizing agents to provide CI/CO services, Safaricom naturally wants to encourage M-Pesa users to spend their e-money within the ecosystem instead of cashing out. Developing M-Pesa's financial ecosystem enables M-Pesa customers to perform more digital transactions, resulting in more transaction revenues for Safaricom, but also less agent commission payments for cash-out, and a further profitability boost. Building the M-Pesa ecosystem to provide opportunities for more non-agent-based transactions increases the velocity of money on the platform.

Table 5.5 Timeline of M-Pesa Expansion of Functionality and Services, 2005–13

Year	*Service name*	*Description*
2005	Sambaza	Airtime sharing
	MFI pilot	Loan repayment tool
2007	M-Pesa	Payment service
2008	PesaPoint ATMs	Cardless ATM withdrawals from M-Pesa account
	Postbank	M-Pesa agent
	Old Mutual	Top-up unit trust accounts
2009	KPLC	Bill pay through M-Pesa for electricity utility
	SMEP	MFI loan repayment through M-Pesa
	KCB	Superagent providing liquidity to M-Pesa agents
	Kenswitch ATMs	Card to any M-Pesa account from 650 ATMs
2010	Equity ATMs	Cardless ATM withdrawals from M-Pesa account
	Musoni	MFI disbursing loans and receiving individual repayments through M-Pesa
	M-KESHO	Equity Bank account accessed through M-Pesa
	Nunua na M-Pesa	Pay at Naivas and Uchumi supermarkets with M-Pesa
2011	Lipa Karo na M-Pesa	Pay school fees with M-Pesa through PesaPal and KCB
	Visa prepaid card	Safaricom and I & M Bank cobranded card topped up directly from M-Pesa
	Western Union	International money transfer from 45 countries directly to M-Pesa account
	M-Bima-Jijenge Savings Plan	Endowment plan from CIC Insurance with daily contributions of K Sh 20 (US$0.23)
	Kilimo Salama	Crop insurance from UAP Insurance
	Mbao	Pension scheme for the informal sector with daily contributions of K Sh 20 (US$0.23)
2012	Linda Jamii	Health insurance
	M-Shwari	Microcredit and microsavings product
2013	Lipa na M-Pesa	Pay merchants with M-Pesa at no charge to customers
	Lipa Kodi	Pay rent with M-Pesa

Source: World Bank.
Note: ATM = automated teller machine; KCB = Kenya Commercial Bank; MFI = microfinance institution.

Merchant Payments

M-Pesa was designed for individual use—which is fine if you are paying your taxi driver, as he just uses his personal account—but how do individual tellers in a shop reconcile M-Pesa with their till? Consumers in Kenya pay US$12.8 billion to other consumers (30 percent of their payments), but they pay US$29 billion to businesses (69 percent of their payments). The balance of US$0.3 billion (1 percent of payments) is paid to the government (Bill and Melinda Gates Foundation 2013). The greater opportunity is therefore in merchant acceptance.

At the end of 2010, Safaricom entered into its first merchant agreement, which would allow customers to pay with M-Pesa at the supermarket chains Naivas and Uchumi. By 2011, however, M-Pesa had only 100 individual stores signed up, and it took an innovative merchant acquirer, Kopo Kopo ("transactions"), to convince Safaricom of the potential opportunity. A huge amount of money is circulating digitally through M-Pesa, but there was no easy way to use it to pay for things. Kopo Kopo managed to convince Safaricom to change its charging structure to launch a new service. Instead of customers paying tiered fees for transactions, merchant payments would be free for customers, with the merchant paying a fee of 1.5 percent. Safaricom signed a merchant aggregator agreement with Kopo Kopo in March 2012. But Safaricom was not satisfied with an arrangement in which only Kopo Kopo acquired merchants on their behalf. They also started aggressively recruiting merchants themselves, directly competing with Kopo Kopo and targeting 100,000 merchants in the next six months.

Lipa na M-Pesa was launched in June 2013. Although Safaricom has a contractual relationship with Kopo Kopo, two months after launch Safaricom dropped the merchant fee to 1 percent without notice, which Kopo Kopo has matched. This 1 percent merchant fee to accept M-Pesa is significantly lower than the 2.5 percent typical for accepting payment cards. In addition, Safaricom has not shared its application programming interface (API) with Kopo Kopo, meaning that any changes on the M-Pesa platform can cause knock-on headaches for Kopo Kopo. Kopo Kopo has recruited 10,000 merchants onto its platform, about a third of the merchants on Lipa na M-Pesa. They are recruiting at a rate of 1,500 per month.

Kopo Kopo enables small and medium enterprises (SMEs) to accept, process, and manage mobile money payments. But, given that it is in direct competition with Safaricom, it has to do more. MasterCard statistics show that 90 percent of small businesses still track payments via pen and paper (Miller and Salazar 2013). Kopo Kopo's platform offers a number of tools online or in a mobile app to help merchants track customers and transaction trends; they also offer real-time settlement to bank accounts.

Following half-year results in 2013, Safaricom stated that it intended to continue growing M-Pesa by growing Lipa na M-Pesa and cashless distribution, e-commerce, transport payments, bank-to-M-Pesa linkages, and uptake of savings and loans products (Safaricom 2013). Safaricom's plans for a near-field communication card management system may indicate its desire to move further still.

PesaPal is another company providing merchant services. Its focus is to enable merchants to accept mobile money, Visa, and MasterCard for online payments. Merchants are charged 2.7 percent of the value of transactions to access these payment channels. PesaPal payments also settle through a bank account, meaning that merchants do not have to be official partners with Safaricom, which settles to M-Pesa. PesaPal has been particularly successful helping schools to accept electronic payments. It is also investigating utilizing mobile point of sale (M-POS) to economically serve small merchants for card payments. PesaPal also offers individual users a PesaPal e-wallet. Users can top up their virtual wallets through PesaPoint ATMs, M-Pesa, Airtel Money, or yuCash and then make payments directly from it.

Another online and mobile payment platform is JamboPay, launched by tech firm Webtribe in 2009. JamboPay now has 700 businesses and over 100,000 users on its platform (Matinde 2013). The JamboPay service provides value to merchants by aggregating multiple payment channels such as M-Pesa, Airtel Money, yuCash, Orange Money, Visa and MasterCard debit and credit cards, as well as direct bank account debiting and crediting. JamboPay was the 2013–14 winner of the Google Innovation Award in the financial sector category.

One of the first service enhancements was the use of ATMs to provide M-Pesa account holders, who are not bank account holders, with 24-7 access to cash. This approach was pioneered in Kenya, but the relatively low stock of ATMs (fewer than 2,500) limited the extension of mobile money. This approach is much more interesting in countries with large numbers of ATMs—for example in Thailand, where more than 50,000 ATMs can be used for mobile money CI/CO (Pénicaud and Katakam 2014). Another service, the bill payment function, was designed as a way for customers to pay their utility bills directly from their phones without having the inconvenience of waiting in long lines. Kenya's electricity utility was the first company to accept bill payment through M-Pesa.

M-Pesa's huge customer base is larger than that of any bank or other financial sector organization in Kenya. As a result, after banks initially criticized the system and even lobbied the Central Bank of Kenya (CBK) to close it down, they started joining M-Pesa. Postbank was the first bank to join M-Pesa as an agent, just five months after launch. Today numerous banks and MFIs are linked to M-Pesa, enabling them to receive funds from customers through the bill-pay function, and smaller numbers of banks have a USSD link that integrates their customers' accounts with M-Pesa, enabling their customers to move money to and from M-Pesa 24 hours a day (table 5.6).

Table 5.6 Banks and MFIs Linked to M-Pesa, 2013

#	*Bank or MFI*	*Bank to M-Pesa (US$)*	*M-Pesa to bank (PayBill)*
Banks			
1	African Banking Corporation (ABC) Bank		✓
2	Bank of Africa	✓	✓
3	Barclays Bank K Ltd.	✓	✓
4	CfC Stanbic	✓	✓
5	Chase Bank	✓	✓
6	Citibank N.A. Kenya		✓
7	Commercial Bank of Africa	✓	✓
8	Consolidated Bank Ltd.	✓	✓
9	Co-operative Bank	✓	✓
10	Diamond Trust Bank	✓	✓
11	Ecobank	✓	✓
12	Equatorial Commercial Bank		✓
13	Equity	✓	✓
14	Family Bank Ltd.	✓	✓
15	First Community Bank Ltd.	✓	✓
16	Gulf African Bank	✓	✓

table continues next page

Table 5.6 Banks and MFIs Linked to M-Pesa, 2013 *(continued)*

#	Bank or MFI	Bank to M-Pesa (US$)	M-Pesa to bank (PayBill)
Banks *(continued)*			
17	Housing Finance Company Ltd.	✓	✓
18	I & M Bank Ltd.	✓	✓
19	Imperial Bank Ltd.		✓
20	Jamii Bora Bank		✓
21	Kenya Commercial Bank	✓	✓
22	K-Rep Bank	✓	✓
23	National Bank	✓	✓
24	National Industrial Credit Bank Ltd.	✓	✓
25	Post Office Savings Bank	✓	✓
26	Prime Bank		✓
27	Standard Chartered Bank	✓	✓
28	Transnational Bank	✓	✓
29	United Bank for Africa		✓
MFIs			
30	Faulu DTM	✓	✓
31	KADET Ltd.	✓	✓
32	KWFT DTM	✓	✓
33	Musoni	✓	✓
34	Rafiki DTM	✓	✓
35	SMEP DTM	✓	✓
36	Uwezo DTM		✓

Source: Safaricom 2013.
Note: DTM = Deposit Taking Microfinance; KWFT = Kenya Women Finance Trust; MFIs = microfinance institutions.

ATM Transactions: Third-Party Providers

PayNet was the first company in the world to initiate cardless ATM transactions when its PesaPoint ATMs became cash-out locations for M-Pesa in July 2008. This provided M-Pesa users with 24-hour access to funds in their mobile money accounts. This is particularly interesting from a financial inclusion point of view because it not only increased the revenue for ATMs but also exposed cardless M-Pesa users to an important piece of mainstream payment infrastructure, increasing their knowledge and understanding of it.

PayNet is a financial services support company that is best known in Kenya for its PesaPoint-branded independent ATM network. PayNet started its independent ATM network in 2005 in an effort to support interconnected services. It has 180 PesaPoint-branded ATMs and 34 banks on its network, interconnecting 800 ATMs. The company aims to interconnect as many financial institutions as possible. Its ATMs provide shared infrastructure for small- and medium-size banks and a backup network for the larger banks. PesaPoint charges just K Sh 30 for ATM transactions.

PayNet also provides a debit and credit card processing center for banks, acts as a third-party issuer for Visa and MasterCard, and runs a POS-enabled agent network selling mobile airtime. The company also offers a wage payment system called Wagepoint that issues closed-loop cards and places ATMs at employers' sites.

This service is primarily used by employers of large numbers of unbanked workers, such as flower and tea farms, that have previously relied on time-consuming cash payment arrangements. Closed-loop cards are also made available to savings and credit cooperatives and MFIs.

Musoni Microfinance; Using M-Pesa as Its Core Banking System

Musoni (M for "mobile" and usoni meaning "future") was established in 2009 to showcase how a cashless and paperless MFI could be transformative. Musoni uses electronic payments through mobile money for all loan repayments and disbursements as well as deposits and withdrawals of savings.

One of the reasons that a pre-M-Pesa pilot with Faulu, a large Kenyan MFI, failed was because of the MFI's concern that using mobile money for loan repayment would undermine its group-lending methodology. Musoni, however, has managed to adopt the technology without undermining its group-lending methodology. Instead, it has used M-Pesa as its core banking system to reduce costs and increase the quality of service to customers. Musoni has effectively outsourced the expense of cash handling to M-Pesa. (Musoni estimates that approximately 50 percent of resources in a typical MFI are consumed by cash handling.) The company's branches have very low costs because they have no safes or security measures in place; branches are merely convergence points for purely administrative purposes. Loan officers are much more efficient than in traditional settings, as they do not have to spend any time banking. Field branch offices are also paperless, having digitized all parts of the application process with the use of tablets.

Since Musoni issued its first loan in May 2010, it has disbursed 18,000 loans totaling K Sh 500 million (US$6 million). Musoni's minimum loan size is US$58, and its average loan size is US$500. It currently serves 8,000 clients and handles 50,000 transactions per month through its proprietary M-Pesa-based core banking system. The company's loan book stands at US$3.5 million, with nonperforming loans at zero. It believes that the quality of its loan book is in part due to the financial discipline provided by using mobile money.

Musoni has a 100 percent repayment record because groups need to pay the day before their loan meeting. Group meetings are a maximum of one hour in length, and this time is used exclusively for training, including business management, public hygiene, current affairs, and introduction of useful products and services, which helps to motivate attendance. Most repayments are after 5 p.m. and on Sunday, when people have time that does not impinge on their daily business activities. Musoni's solution allows the disbursement of loans within 24 hours of application directly to a borrower's M-Pesa account. It also provides privacy in loan repayment while being supported by the group guarantee.

M-Shwari: Savings and Loans

M-Shwari is a bank account, available only through M-Pesa, that can be used for savings as well as a means to access microcredit. M-Shwari was launched in November 2012. It is a product of the partnership between Safaricom and CBA, although most customers are unaware of the bank's involvement.

Safaricom already had a long-standing relationship with CBA: as a corporate client, CBA held the first M-Pesa trust account. (CBK still holds an M-Pesa trust account, but funds are now split among three banks.) CBA is a bank for corporate customers and high-net-worth individuals with a strong focus on technology. The success of M-Shwari has boosted its customer base from just 37,000 to 5 million, the second largest bank customer base behind Equity Bank.

Since its launch, M-Shwari has reached nearly 9.2 million customers, with deposits of almost US$1.5 billion and loans of US$277.2 million (figure 5.14 and table 5.7). The speed of uptake is remarkable, possibly indicating a large unserved market. M-Shwari loans are available in a real-time credit to an individual's M-Pesa account. To qualify for a loan, an applicant must have been an M-Pesa subscriber for at least six months, and then an algorithm based on their use of M-Pesa and airtime purchases is used to determine the initial loan limit.

Figure 5.14 Growth in Number of M-Shwari Savings Accounts, 2013–14

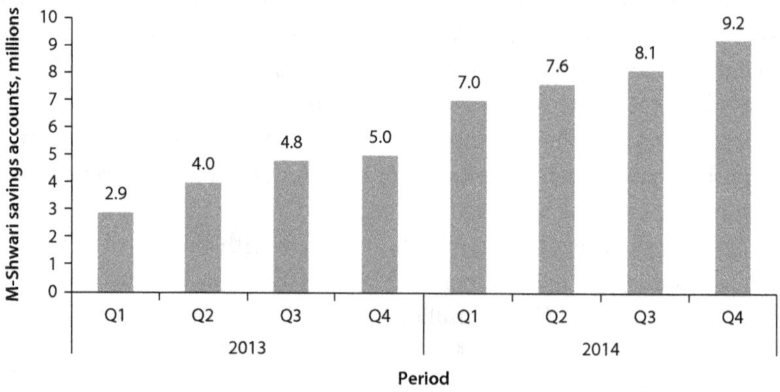

Sources: CBA 2013; Cook and McKay 2015.

Table 5.7 Key M-Shwari Statistics

Measurement	*Statistics*
Total savings accounts	9.2 million opened 7.2 million unique customers 4.7 million active 90 days
Total number of loans	20.6 million cumulative loans since launch 2.8 million unique borrowers since launch 1.8 million active as of December 2014
Deposit amounts	US$1.5 billion deposited since launch US$45.3 million deposit balance as of December 2014
Loan amounts	US$277.2 million disbursed since launch US$17.7 million outstanding as of December 2014
Average savings balance	K Sh 504 (US$5.56)—all accounts K Sh 911 (US$10.06)—active 90 days K Sh 1,971 (US$21.76)—active 30 days
Nonperforming loans	2.2 percent over 90 days

Sources: CBA 2013; Cook and McKay 2015.
Note: K Sh = Kenya shillings.

This microloan product is serving the base of the pyramid with a minimum loan size of just K Sh 100 (US$1.15), a maximum loan size of K Sh 20,000 (US$235), and an average loan size of K Sh 305 (US$3.50). Loans are for 30 days and have a 7.5 percent fee. The majority of loans (70 percent) are repaid before the one-month due date. Backers suspect that this is an indication of the importance of this line of credit to individuals. Subsequent loan limits are based on the levels of regular savings with M-Shwari and customers' repayment history on the initial loan.

These customers are below the threshold of most MFIs. M-Shwari is a very different product from a typical MFI loan, which takes a long time to obtain while a group is formed and savings are accumulated as security against a loan. In addition, this product provides a confidential service on an individual basis.

CBA spent two years in product development for M-Shwari, using data from the FinAccess Survey and receiving some assistance from Financial Sector Deepening (FSD) Kenya (FSD Kenya and CBK 2013). The survey showed that people believe savings should be free and held with a trusted person or institution. Focus group discussions revealed that individuals want their bank both close and far. M-Pesa, people said, made it too easy to access one's money. They needed structure to help them with savings discipline, but they also wanted flexibility.

CBA uses the digital footprint that users develop through their use of M-Pesa and airtime purchases to prescore them for credit. This allows CBA to offer a truly unsecured loan. CBA then makes a real-time query of the National Population Register to verify identity. (CBA is the first nongovernment user of the database.) CBA is happy to remain in the background of this product, given the reputation of Safaricom versus the banks in Kenya. The aim was for the product to break even in a little under two years, but it broke even in just eight months.

Multi Media Mobile: An App that Facilitates International Remittances

Multi Media Mobile (MMMobile) is a software company that provides interoperability between the card and mobile money ecosystems by facilitating card-to-M-Pesa transactions through an app on a user's phone. The company believes that cards will form the predominant digital payment technology in the future. However, at the moment, mobile money predominates in Kenya. Hence the company's approach is to build trust in cards through M-Pesa. The SMS and USSD platforms used by mobile money cannot support the level of security associated with card encryption, yet the public in Kenya trusts M-Pesa over cards.

The company's initial target market is for international remitters, who link the card they want to use to the MMMobile app, allowing them to send money to any mobile money account in Kenya. They are also targeting tourists who want to buy from smaller merchants who do not have card acceptance devices; the MMMobile app allows them to still have the security of not carrying cash while allowing them to pay through mobile money, which is more readily accepted.

This innovation has the potential to trickle down to lower-income users. As with M-Pesa, lower-income individuals' first experience of the system may be by receiving funds from a friend or family member. MMMobile again broadens individuals' remittance ecosystem to cardholders based anywhere in the world. It is particularly interesting to note that it is not a requirement to hold a bank account for either a sender or a recipient. A sender could transfer funds from their prepaid account to a mobile money account and in the future from card to card.

Additional Important Lifestyle Services for the Bottom of the Pyramid

M-Bima: Savings and Life Insurance. In 2011 the CIC Insurance Group, the third largest insurance company in Kenya, introduced a new technology platform called M-Bima ("mobile insurance" in Kiswahili) to strengthen the scale and efficiency of its microinsurance operations. The platform allows customers to send their premiums via M-Pesa and Airtel Money, through the PayBill function. CIC does not charge its customers for transfers into the M-Bima account and even bears the Safaricom charges of K Sh 30 for each transfer.

The first product, launched in 2011 on the M-Bima platform, was the Jijenge Savings Plan. Clients can save as little as K Sh 20 (US$0.23) a day using M-Pesa or Airtel Money and receive SMS reminders to stimulate savings. The product provides clients with a convenient and safe way to build savings. It is a 12-year endowment plan with monthly installments of a minimum K Sh 600 (US$7) for a minimum coverage of K Sh 50,000 (US$580). The product combines savings with life insurance. Immediate coverage is provided for accidental death and, after three months, for natural death. An exit benefit is available at the end of the third year, with a surrender value of K Sh 20,000. Customers can cash out their savings from the third year but can also take the money as a loan from their own account. The target customers are those employed in the informal sector with low and erratic incomes. The product was developed with assistance from the Microinsurance Innovation Facility, housed at the International Labour Organization's Social Finance Programme.

Kilimo Salama Plus: Crop Insurance. In February 2011 UAP Insurance and the Syngenta Foundation extended their insurance product for smallholder farmers with Safaricom and M-Pesa. The low-cost mobile-phone payment and data system is linked to automated, solar-powered weather stations to issue an insurance policy and rapidly compensate farmers for investments in seeds, fertilizer, and other inputs that are lost to either drought or flood. Kilimo Salama ("safe farming") is the largest agricultural insurance program in Africa and the first to use mobile-phone technology to speed access and payouts to rural farmers. Farmers purchase Kilimo Salama Plus through local agrodealers, who use a camera phone to scan a special bar code that sends the policy to UAP over Safaricom's mobile data network. Premiums are collected at the store and sent via M-Pesa to UAP.

This index-based weather insurance relies on the data collected and relayed from 30 weather stations. When data from a particular station indicate rainfall is either 15 percent above or below historical averages and hence likely to significantly reduce crop yields, all farmers with a policy registered with that station automatically receive payouts directly to their M-Pesa accounts at the end of the growing season. Kilimo Salama is supported by the International Finance

Corporation, the Global Index Insurance Facility (which is supported by the European Commission), and Syngenta Foundation for Sustainable Agriculture (Syngenta Foundation 2011).

Mbao Pension Scheme. The Mbao pension scheme targets the estimated 8.9 million informal workers in various sectors of the economy. At the press launch of the scheme in June 2011, the chairman of the Retirement Benefits Authority looked forward to a scenario in which beneficiaries were provided for in old age and funds were raised for investment in the Kenyan economy. Unfortunately, although contributions may be received electronically, a physical application form must be collected from Mbao chapter offices, and pension payments currently have to be made by check. The scheme is gaining in popularity, and by the end of 2012 membership stood at 40,000 (RBA 2013).

"Mbao" refers to the K Sh 20 note (US$0.23), which is the daily minimum contribution required by this individual pension plan. The Kenya National Jua Kali Cooperative Society developed this defined contribution scheme for the informal sector. The project started with a nine-month pilot project in which 11,000 informal sector workers made contributions through the bill payment function on M-Pesa or Airtel Money. Any Kenyan over the age of 18 years is eligible to join this individual pension plan. To join, a membership fee of K Sh 100 is paid, and a commitment is made to contribute at least K Sh 100 per week. The fund is run professionally with a conservative investment strategy focused on investment in money market and fixed income securities. The KCB acts as custodian, the Co-op Trust part of the Cooperative Bank invests the funds, and administration is by Eagle Africa Insurance Brokers.

Linda Jamii: Health Insurance. "Linda Jamii," Swahili for "protect the family," is a private medical care insurance that uses M-Pesa to collect premiums and to make benefit payouts. In November 2012 Changamka Microhealth and Britam Insurance Company launched the mechanism, which allows people to save small amounts of money over time toward purchasing health insurance. They started a pilot with Safaricom in May 2013 and have 2,000 households enrolled to date. Customers can save in a premium deposit facility on M-Pesa and purchase health insurance once they accumulate K Sh 6,000. They then have an additional six months to pay the outstanding K Sh 6,000.

Changamka's strategies to address barriers to coverage include exclusive savings toward health care expenses, flexible timing of payments, affordable access to plans, and use of available mobile technology. In addition, the microinsurance program benefits health care providers, because the electronic operating platform reduces the administrative burden on hospitals and clinics. The product targets the estimated 35 million uninsured in Kenya. Changamka plans to scale up this initiative nationally (Gathara 2013).

To subscribe to the service, users dial *525# and follow the directions. After subscription there is a requirement to complete the registration process by visiting the nearest Linda Jamii agent outlet at Britam offices, Safaricom shops, selected M-Pesa agents, Uchumi outlets, or Postbank branches countrywide. Once a person is registered, a unique Linda Jamii number is used to make payments through Safaricom's PayBill function.

M-POS

The growth of alternative delivery channels to provide financial services has been called "branchless banking," because it relies on the use of channels other than traditional bank branches. Using these alternative channels holds the prospect of extending access to financial services because of the potential for a dramatic reduction in the cost of provision.

In addition to mobile money, another exciting emerging technology is being used in Kenya: M-POS, a card reader attached to a smartphone creating a low-cost POS center for use by smaller merchants. Because of its lower cost, it has the potential to massively increase card acceptance infrastructure and hence card use. (See figure 5.15 for a comparison of costs for various channels.)

Challenges to M-Pesa

Although M-Pesa is truly inspiring innovators in Kenya and globally to develop apps to provide services using the M-Pesa payment platform as the digital connector, the biggest complaint from app developers is the reluctance of Safaricom to share the M-Pesa API with them. API is the technology that allows two software programs to communicate with each other and allows any potential service developer to get linked directly to the mobile money account in M-Pesa instead of a bank account or credit card for payment.

With more than 12 million active accounts, it is no surprise why all developers want to have access to M-Pesa. Everyone in the application developer

Figure 5.15 Average Capital Expenditure Costs for Financial Service Providers in Kenya, by Channel

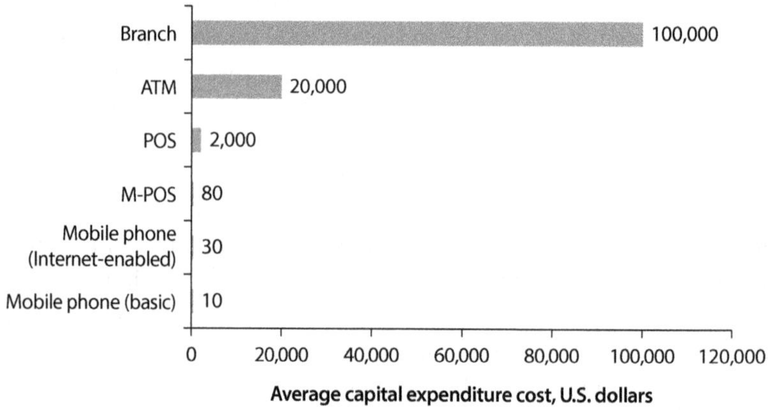

Sources: Branch, ATM, and POS costs from FSD Kenya 2013; M-POS and mobile-phone costs from expert interviews in November 2013.
Note: ATM = automated teller machine; M-POS = mobile point of sale; POS = point of sale.

community has been waiting for ages for Safaricom to release an M-Pesa API that would enable them to build innovative, value-added services on the M-Pesa platform or to link up already developed apps. In 2015, Safaricom took initial steps to open its API to the public to help developers build platforms that can use M-Pesa for quick payments. However, this promise has been there for some time, and the delay had led to independent development of open-source M-Pesa APIs. However, "the systems are not yet in place to facilitate seamless integration into Safaricom's platforms," and opening up would entail technical, financial, risk management, and market implications for M-Pesa to consider (Morawczynski 2015). Nevertheless, opening up the M-Pesa API could be beneficial for the communities that otherwise do not get served.

Digitizing Social Grant Disbursement Programs: Brazil, Mexico, and South Africa

> *In a number of countries, two separate, but potentially complementary policy agendas have emerged in the past five years: governments have sought to increase the use of electronic means for government payments and to promote greater financial inclusion. While the two agendas have by no means converged yet, in practice they have often been translated into a single headline objective: to increase the proportion of recipients of government social cash transfers who receive payment directly into a bank account.*
>
> —Chris Bold, David Porteous, and Sarah Rotman, "Social Cash Transfers and Financial Inclusion: Evidence from Four Countries"

According to the 2014 United Nations E-government Survey, public administration can become more efficient and widely used through the use of recent digital developments, such as mobile apps and innovative financial instruments (UN 2014). The survey report identifies three types of e-government interactions: government-to-government (G2G), government-to-business (G2B), and government-to-consumers (G2C).[15] This volume focuses on reaching the underserved through G2C (or, alternatively, G2P) methods.

National governments and the international community are increasingly recognizing the value of social transfers (including pensions, grants for families, public works schemes, and other programs) in achieving the Millennium Development Goals. In addition to their vital social contribution, social transfers can support critical economic objectives. Many of the world's fastest-growing economies over the past several decades have built social protection programs into their policies at early stages because of their potential to increase productivity and help stabilize domestic demand (Samson, Mac Quene, and van Niekerk 2006).

Social grant payments are disbursed as cash payments in most developing countries. Barriers to transitioning such payments to digital systems include inadequacy of payment infrastructure outside of urban areas, absence of proper identification documents, lack of awareness and trust in payment options and

Figure 5.16 Share of Adults Receiving Government Transfers, by Region and Payment Method, 2014

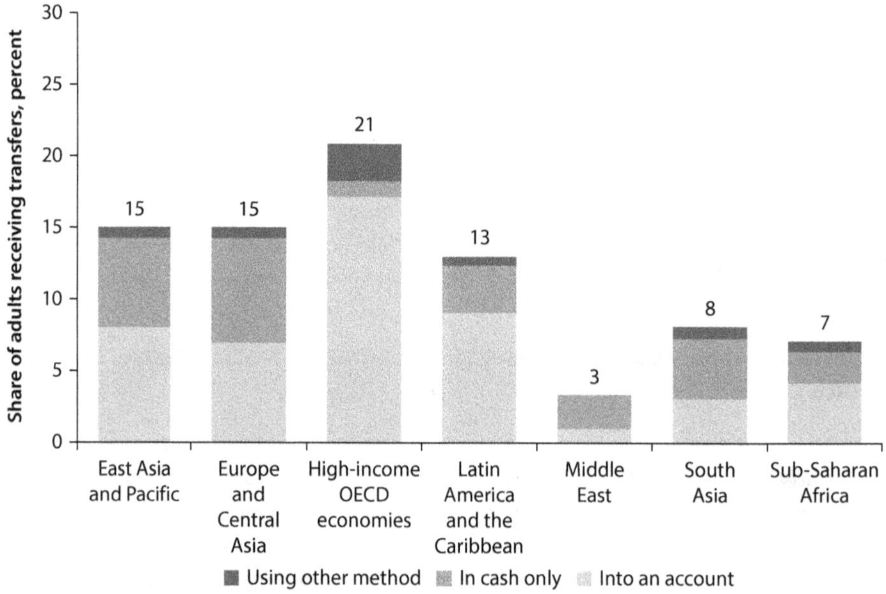

Source: Global Findex Survey 2014, http://datatopics.worldbank.org/financialinclusion/.
Note: OECD = Organisation for Economic Co-operation and Development. Figure displays percentages of adults who reported receiving transfers over the previous year.

institutions, cost of transition, regulatory barriers, and lack of preparedness by governments for digitization. According to Findex 2014, whereas in high-income Organisation for Economic Co-operation and Development (OECD) economies a majority of those receiving transfer payments (83 percent) receive them into an account, in developing countries only about half of them reported receiving such payments into an account (figure 5.16).

In many developing economies, government grant disbursement programs are used to ensure opening up of bank accounts, thus enhancing financial inclusion. Findex 2014 reported that around 61 percent of grant recipient adults also use these accounts for cash management purposes. Four countries in particular—Brazil, Colombia, Mexico, and South Africa—use digital means in various forms to disburse grant payments on a scale that is significant in terms of financial inclusion. While all four use digitized payments, they adopted different methods (table 5.8).

The grant payment schemes from Brazil and Mexico are discussed briefly, highlighting the magnitude of these programs and the opportunities in terms of sheer numbers of people who can be brought into the mainstream financial system. However, because of the proximity issues of formal financial institutions, both programs still remain largely transaction- or cash-based. The South Africa case, on the other hand, highlights superior technology with important features that have supported financial inclusion in a more meaningful manner.

Table 5.8 Payment Approaches of Selected Grant Programs as of 2012

Program aspect	*Brazil*	*Colombia*	*Mexico*	*South Africa*
Frequency of payment	Monthly	Bimonthly	Bimonthly	Monthly
Payment provider	Caixa Economica Federal (state-owned bank)	Banco Agrario (state-owned bank)	Bansefi (state-owned bank)[a]	Any bank or specific payment providers in different provinces: Net 1, Empilweni, and AllPay (subsidiary of ABSA Bank)
Physical cash	1%	9%	66%	0%
Limited-purpose instrument	84% Magnetic-stripe debit card (Social Card) whereby funds must be withdrawn within 60 days at Caixa agents or various ATM networks and no additional funds can be deposited	91% Magnetic-stripe debit card that can be used only at Assenda merchants and ATMs	0%	41% Specific payment providers that offer store-of-value via smart cards, but no additional fund deposits and use only at dedicated pay points
Mainstream financial account	15% Caixa Facil basic bank account with magnetic-stripe card	0%	34% • 16%—Bansefi savings account with magnetic-stripe card (Debicuenta) • 12%—Bansefi prepaid account with smart card • 6%—Bansefi passbook account[b]	59% Mainstream bank accounts; Sekulula account (offered by AllPay) with magnetic-stripe card as default option in certain provinces

Source: Bold, Porteous, and Rotman 2012.
Note: ATM = automated teller machine.
a. In 2011, state communications agency Telecomm was still involved as a direct payment provider. But Oportunidades is in the process of consolidating its payments previously made through several agencies, including Telecomm, so that all payments would be made in 2012 through Bansefi, which then subcontracts with other networks, including Telecomm, to effect payout.
b. These cash payments into accounts are being phased out.

Brazil: Bolsa Família

From 2001 to 2003 Brazil created four cash transfer programs that were merged into the Bolsa Família program (BFP) in late 2003 and subsequently expanded. The new program seeks to invest in human capital by associating cash transfers with educational goals and uptake of health services (Paes-Sousa, Santos, and Miazaki 2011). BFP is the world's largest conditional cash transfer program, reaching all 5,564 municipalities in the 27 states of Brazil. Over 12.9 million households are enrolled, benefiting over 60 million people (about 30 percent of the total population).

Bolsa Família owes much of its targeting success to Cadastro Único (CadÚnico), a single registry that has consolidated data on Brazil's most vulnerable and has enabled Brazil to expand and improve its social welfare system.[16] CadÚnico is managed by Caixa Economica Federal (Caixa), a state bank.[17] Eligibilities are redetermined every two years. Local governments collect the data and the information, while the federal government, through the Ministry of Social Development, oversees the registry, defines eligibility requirements, and cross-checks data within the registry. Caixa plays a distinct role in implementing BFP by managing the database, assigning identification numbers, and finally distributing grants.

In terms of payment method, around 1 percent of the recipients receive cash, while 15 percent receive the grants through basic bank accounts held with Caixa. These are no-frills accounts, easier and faster to open by regulation, and offering four withdrawals free each month. Monthly balances are limited to R$2,000 (US$575). A majority of the recipients, around 84 percent, get their grants using a limited-purpose card—the Social Card. This card is linked to a nontransactional account that requires withdrawal of the grant within 90 days in one transaction from the day of the deposit and does not allow other deposits. There is a nominal fee for the withdrawal. If money is not withdrawn, the grant recipient will lose that particular payment—the reason why the Findex 2014 Survey found that close to 88 percent of grant recipients withdraw all the money right away.

Caixa has a very effective network of payment agents of over 36,000 points of service in all municipalities of the country, including 2,780 branches, 24,756 retail agents, and 10,954 lottery outlets (CGAP 2011). A majority of the recipients (around 61 percent) withdraw money at lottery centers. Caixa is conducting pilot studies on collecting biometrics, use of mobile phones and enhancing the touchpoints. Efficient administration and good targeting have enabled BFP to achieve its success at a very low cost.

Ten years after its inception, BFP has been key to helping Brazil cut its extreme poverty by more than half—from 9.7 percent of the population in 2003 to 4.3 percent in 2013. Most impressive, and in contrast to other countries, income inequality also fell markedly, by 15 percent, to a Gini coefficient of 0.527 (Wetzel 2013). Given the structure of the payment process, the Social Card remains a transactions card and not a savings mechanism for most grant recipients. However, efficient targeting by the CadÚnico registry, wide distribution of agent touchpoints, and effective management of the whole process have made BFP a powerful tool in fighting poverty.

Mexico: Oportunidades (Prospera)

The conditional cash transfer scheme in Mexico was originally founded in 1989 and was rebranded twice before being renamed as Prospera in 2014.[18] This was the first major social grant program in Latin America where grants were conditioned on activities such as school enrollments and getting regular health checkups. To date, this model has been replicated in as many as 52 countries. The program serviced around 5.8 million families, which is around 20 percent of the population in 2011, and in 2014 this increased to 6.1 million families.

The feeling in Mexico had been that Oportunidades was failing to make a meaningful long-term impact on the country's social panorama, with atrocious levels of poverty and inequality refusing to budge over the years. This led the government to relaunch a widened program under the new name of Prospera, in the hope that it would finally reverse a trend of stagnant and widespread poverty across Latin America's second biggest country in terms of the size of its population and economy (Constantine 2014).

Prospera is also promoting beneficiaries' access to higher education and formal employment. Additionally, Prospera is facilitating access to financial services (savings, microcredit, and insurance), thereby enhancing social inclusion of the country's poorest citizens (World Bank 2014). The Social Development Secretariat (SEDESOL) is working to develop an integrated social information system to identify the poor, and it hopes to have a system similar to the Cadastro Único in Brazil.

The payments under Oportunidades were made bimonthly through the state-owned development bank, Bansefi. However, 66 percent of payments were made in cash. The rest were distributed through Bansefi accounts: 16 percent through savings accounts with magnetic-stripe cards; 12 percent through Bansefi prepaid accounts operated through smart cards; and the remaining 6 percent through Bansefi passbook accounts. In 2012, the amount transferred to 6.5 million beneficiaries was Mex$63.78 billion (US$3.9 billion). Given that most beneficiaries live in rural, hard-to-reach areas, SEDESOL has been continually trying out various new payment partners and options—Telecomm (the state telegraph company), gas stations, cooperatives, and Diconsa stores,[19] to name a few.

In 2010 the government mandated SEDESOL to centralize the electronic payments of the grant program. As a result, all payments were outsourced to Bansefi, with Telecomm and Diconsa as subcontractors. Where banking infrastructure is available, the beneficiaries are issued a no-fee open debit card to be used at any ATM or POS terminal,[20] and the rest are issued a biometric closed debit card that can cash out only at Bansefi, Telecomm, or Diconsa touchpoints. Even though Diconsa stores have the physical outreach, the National Banking and Securities Commission approved fewer than 290 stores as banking agents. Hence this method, too, failed to gain traction. By 2012, only 20 percent went through open cards, with the rest paid in cash at either fixed points (20 percent) or at temporary points (61 percent) (Babatz 2013).

The lesson is that mandatory regulatory action alone will not bring about the desired results, unless backed by a well-designed strategic plan. Furthermore, proximity to payment touchpoints is especially important in programs that serve the rural poor. Transaction cost is not the only thing poor people worry about; hidden opportunity costs of travel, loss of daily wages, or even the safety of carrying cash are all factored into the decision of choosing payment alternatives. The positive impacts of this conditional cash transfer program can be fully realized by addressing these issues, possibly through innovative use of digital solutions.

South Africa: Card-Based Biometric Grant Payment Disbursement System

Millions of South Africans lack access to the most basic financial tools. They don't have secure places to save money or reliable means to transfer it and use it for transacting. Through the introduction of the SASSA Debit MasterCard card, nearly one fifth of the South African population now benefits from having a formal banking product, helping them build a stronger future for themselves, their families and their communities.

—Ann Cairns, president of international markets, MasterCard, during award presentation to the South African Social Security Agency (SASSA), August 20, 2013

South Africa has seven major social security grant programs: Old Age, War Veteran's, Disability, Grant in Aid, Child Support, Foster Child, and Care Dependency. Eligibility for each grant is dependent on an income-based means test. Administered by the South African Social Security Agency (SASSA), the grants are financed through general tax revenues collected on a national basis. As of June 2015, there were 16.78 million grant recipients (about 31 percent of the total population). Over the five years from 2009–13, social grants were around 3.4 percent of the GDP, and this trend is expected to continue (National Treasury, South Africa 2013).

Social Grants and Financial Inclusion

South Africa is fairly financially inclusive, according to the Findex 2014 Survey: around 70 percent of the adults over 15 years of age have accounts in formal financial institutions.[21] However, the share of adults who save and borrow (33 percent and 12 percent, respectively) leaves room for improvement.

The impact of social grants on financial inclusion is highlighted in the FinScope 2014 Survey (figure 5.17), which shows that 80 percent of the adult population is formally served (through either banks or formal nonbank institutions) while 75 percent is banked. Another 6 percent is served by informal means, while 14 percent of the population is financially excluded. Financial inclusion grew from 17.7 million adults in 2004 to 31.4 million in 2014, and the adults who rely only on informal mechanisms have been reduced from 3.4 million in 2004 to 2.1 million in 2014. In terms of gender, women are more likely to have bank accounts (79 percent) than men (70 percent) and less likely to be financially excluded (11 percent and 18 percent, respectively).

According to the FinScope Survey, the increase in financial inclusion is largely driven by the SASSA grants that are disbursed through a bank account–linked debit card. Accordingly, 34 percent of the banked population now owns a SASSA MasterCard. This is a remarkable achievement. Deeper examination reveals that it is not just the different types of social grants that made such inclusion possible; the technology behind delivering grant payments to 16 million people has helped to bring them into the formal financial system.

SASSA Card: Biometric Chip Technology–Enabled Debit MasterCard

Until 2004, the grants had been allocated to provinces as block grants and disbursed mainly through private companies contracted for this purpose. In 2004, the National Treasury revised this practice and started disbursements through the Department of Social Development, and the payment options offered were cash at specific pay points or direct bank credits. Given the low banking access possibilities for the poor and the high cost of banking, in reality most grants were paid in cash.

Figure 5.17 Financial Access Strand in South Africa, 2004–14

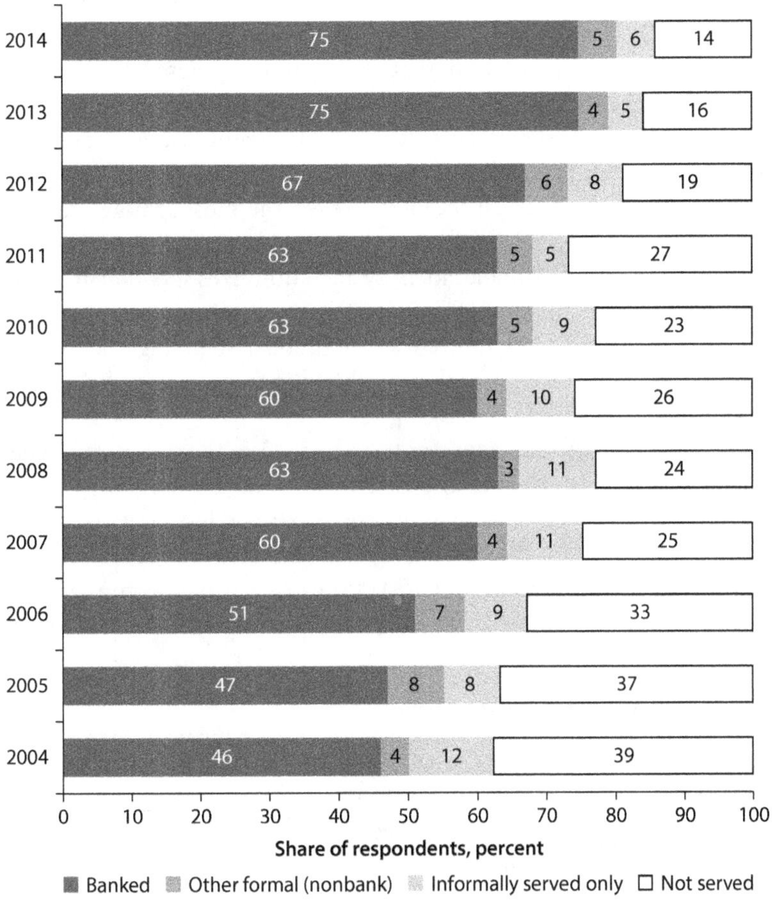

Source: FinScope 2014.
Note: In constructing these strands, the overlaps in financial product or services usage are removed. The figure reflects FinAccess 2014 Survey respondents who were ages 18 and older. The "Access Strand" is a graphical method (routinely used in FinScope Surveys) of placing each survey respondent along a continuum of access, depending on use of services by category. "Formal" refers to the share of the population using a bank, banklike institution, Postbank, or an insurance product. "Other formal (nonbank)" refers to the share using semiformal services from nonbank financial institutions (such as microfinance institutions), not bank services. "Informally served" refers to those using only informal financial service providers such as savings and credit associations and groups or individuals other than family and friends. "Not served" refers to those using no institutionalized financial services.

In 2011, realizing the challenges by a variety of resource and logistical limitations that made grant delivery vulnerable to widespread waste, fraud, and abuse, SASSA contracted Cash Paymaster Services (CPS, a subsidiary of Net 1/ Aplitec), in partnership with Grindrod Bank, to handle the grant disbursements. Grindrod Bank turned to MasterCard as its payment partner of choice. As the designated issuer, Grindrod Bank was able to hold deposits and, as a member of the Payments Association of South Africa, is able to clear through the National Payment System.

Net 1 and MasterCard collaborated on developing a unique MasterCard, integrating Net 1's Universal Electronic Payment System (UEPS) biometric technology with MasterCard Europay, MasterCard, and Visa (EMV) chip technology[22] to create a solution that successfully operated both online and offline,[23] unlike traditional payment systems offered by major banking institutions that require immediate access through a communications network to a centralized computer. This offline capability allows the user to conduct transactions at any time with other cardholders in even the most remote areas, so long as a smart card reader, which is often portable and battery powered, is available. The card authenticates cardholder identity and authorizes spending using a single chip that can be instantly issued. The hybrid chip solution incorporates biometric identification and loading of funds with traditional spending and fund access functionality through e-wallets. Table 5.9 highlights the salient features of this unique and innovative card.

With the launch of the new card, SASSA also took the unprecedented difficult task of reregistering all the beneficiaries of the seven types of social grants. The new technology enabled the registration and instant issuance of a card in just under 10 minutes, thus saving registration and wait times for the grant beneficiaries. In peak times, around 150,000 cards per day were registered and issued. Since the card launch in March 2012, just under 22 million social grant

Table 5.9 Characteristics of the SASSA Card

Category or feature	*Description*
International card scheme	MasterCard
Issuing bank	Grindrod Bank Ltd.
Government agency (cobranding partner)	SASSA
Provider of funding	National Treasury of South Africa
System operator or service provider	CPS/Net 1/Aplitec
Account holders or cardholders	Approved and registered government grant recipients
Card technology	Contains an EMV-compliant chip, which allows fingerprint biometric verification, PIN verification, and debit card purchases. Also contains an offline UEPS Net 1 electronic purse (wallet)
Proof of life authentication	Fingerprint biometric (Net 1 proprietary); voice recognition
Payment transaction cardholder verification method	Fingerprint biometric; PIN
Registration access points	SASSA regional offices
Proof of life access points	SASSA regional offices, selected retailers
Transactional access points	CPS cash payout points (biometric cards); selected retailers (biometrics); retailers (PIN); any ATM (PIN)
Frequency of payments	Monthly

Source: Volker 2013.
Note: ATM = automated teller machine; CPS = Cash Paymaster Services; EMV = Europay, MasterCard, and Visa; PIN = personal identification number; SASSA = South African Social Security Agency; UEPS = Universal Electronic Payment System.

beneficiaries have reregistered in the new system. With 10–12 million SASSA cards issued, fraudulent grant applications have been minimized and administration costs reduced by distributing all grant payments electronically.[24]

As part of the SASSA reregistration process, each recipient has a bank account opened for them, which is offered free of monthly charges by Grindrod Bank. Recipients can deposit funds into their bank account via electronic funds transfer (EFT) or third-party bank transfer. The SASSA Debit MasterCard can be used anywhere MasterCard is accepted, and grant recipients can use it like a normal debit card to make purchases, check their account balances, and withdraw cash without incurring transaction charges at selected South African retailers. Recipients can also withdraw cash at any ATM, although normal banking charges apply.

According to Net 1, unlike a traditional credit or debit card where the operation of the account occurs on a centralized computer, each of the SASSA smart cards effectively operates as an individual bank account for all types of transactions. Although the SASSA card is a single-wallet system, it is possible to enable multiple e-wallets on a single card, where each wallet can be tied to specific payment partners. For example, education grants could only be activated when used to pay school fees. This could potentially enhance cash management capabilities of the beneficiaries and ensure grants are used for the intended purpose.

Future of Grant Payments in South Africa

There is no question that using a unique biometric-enabled UEPS EMV SASSA card transformed the distribution of social security benefits in South Africa. The gains made by reregistration and incorporation of biometrics in the process are substantial. While creating a comprehensive national social grants database, it also helped to reduce waste, abuse, and fraud. Consolidating seven social programs and combining all payments under one card also brought in efficiency gains. According to the CEO of SASSA, the elimination of ghost accounts, duplications, and other irregularities resulted in over 650,000 accounts being removed from the grants register, thus creating annual savings of R 2 billion (US$157 million) (SASSA 2013). An additional R 800 million (US$63 million) in service fees were saved in the 2012/13 financial year alone (Brand South Africa 2014). For the grant beneficiaries, absence of long lines at payment locations with scheduled payment times, reduction of personal security risks of carrying money, prevention of abuse and fraud, increased efficiency and safety, and availability of free transactions and withdrawals are some of the important benefits that enhanced their value proposition.

Nevertheless, it must be mentioned that the SASSA operation is laden with controversies, court cases, and allegations against most stakeholders, including SASSA itself. However, this case study focuses only on the important technical solution offered. Digitizing manual payments is a difficult process, and digitizing seven grant programs and creating one innovative, easy-to-use, bank account–linked debit card is an impressive feat that significantly enhanced the value proposition for the poorest and most vulnerable in terms of their financial inclusion.

Notes

1. The questionnaire was designed to capture innovations resulting in new products as well as innovations in processing. A total of 101 central banks completed the survey and reported 173 innovative retail payment products or product groups. Many central banks provided information on a product group basis and not individual products (see World Bank 2012a).
2. If providers have proprietary technical and usage standards, interoperability would entail higher cost of compliance.
3. For all Global Findex Survey 2014 data, see http://datatopics.worldbank.org/financialinclusion/. For the Findex findings on Sri Lanka, see appendix A, table A.6.
4. KPMG is a leading audit and adviser services provider.
5. Sri Lanka also has the world's lowest rates for fixed broadband Internet, around US$5 per month, resulting from a high level of competition among the suppliers and an explicit aim from the government to keep the cost to a minimum (ITU 2012).
6. The global mobile industry converges on Barcelona annually for the Mobile World Congress, with the 2015 edition attracting a record 93,000 delegates and over 2,000 exhibitors. For more information, see *DailyFT* (2015).
7. Standard setting can also suppress the incentives to innovate and, hence, sometimes can restrict competition by curbing such innovative business models.
8. 1LINK is the leading shared ATM network of Pakistan.
9. Global Findex Survey 2014, http://datatopics.worldbank.org/financialinclusion/.
10. The team was supported by the International Finance Corporation (IFC) of the World Bank Group, the Bill and Melinda Gates Foundation, the Financial Sector Deepening (FSD) Trust, and Groupe Speciale Mobile Association (GSMA) (GSMA 2014).
11. For all Global Findex Survey 2014 data, see http://datatopics.worldbank.org/financialinclusion/.
12. According to the Findex data, the entire 36 percent have access through formal bank accounts and only 0.4 percent through mobile accounts. (Note: there may be sampling issues underlying these numbers.) For more of the Findex 2014 data on Indonesia, see appendix A, table A.2.
13. Indonesia's six current mobile money service providers and scheme names include Telkomsel: T-Cash (begun in 2007); Indosat (Ooredoo): Dompetku (2008); XL (Axiata): XL Tunai (2012); mCoin: mCoin (2012); BTPN (Bank): Wow! (2013); and Bank Mandiri: e-Cash (2013).
14. M-Pesa revenue and expenditure data from Safaricom and authors' field interviews.
15. G2C is also called government-to-persons (G2P).
16. The registry contains data on over 23 million low-income families and 78 million people. Estimates based on the 2010 census data show that there are 20 million low-income families in Brazil (67 million people), or 35 percent of the total Brazilian population. Therefore, there is 114.5 percent coverage. Its biggest program is the BFP. This is a huge undertaking, given that Brazil is the fifth largest country in the world (8.5 million square kilometers) (Mostafa 2014).
17. Established in 1861, Caixa is a government savings bank.
18. It started as Pronasol in 1989, and was renamed Progresa in 1997 and Oportunidades in 2002.

19. Diconsa stores are small stores owned and operated by local communities; there are over 20,000 of such stores, mostly in rural Mexico.
20. Fees are paid by Bansefi.
21. For all Global Findex Survey 2014 data, see http://datatopics.worldbank.org/financialinclusion/. For the Findex findings on South Africa, see appendix A, table A.5.
22. EMV stands for Europay, MasterCard, and Visa—the three companies that originally created the standard. EMV is a technical standard for smart payment cards and for payment terminals and ATMs that can accept them. EMV cards are smart cards (also called chip cards or IC cards) that store their data on integrated circuits rather than magnetic stripes, although many EMV cards also have stripes for backward compatibility. They can be contact cards, which must be physically inserted (or "dipped") into a reader, or contactless cards, which can be read over a short distance using radio-frequency identification technology. Payment cards that comply with the EMV standard are often called chip-and-PIN or chip-and-signature cards, depending on the exact authentication methods required to use them.
23. Real-time but offline.
24. Because many of the beneficiaries are underage or too old to collect funds, collection can be delegated to dependents or nominees, who in turn may be grant beneficiaries. Multiple grant beneficiaries can be loaded onto a single SASSA Debit MasterCard.

Bibliography

Ackerman, C. 2015. "Cadastro Unico: Behind the Scenes of Bolsa Familia." Newsletter, Gerald R. Ford School of Public Policy, University of Michigan. https://iedpbrazil.wordpress.com/2015/02/26/cadastro-unico-behind-the-scenes-of-bolsa-familia/.

Babatz, G. 2013. "Sustained Effort, Saving Billions: Lessons from the Mexican Government's Shift to Electronic Payments." Case study, Better Than Cash Alliance, Somerville, MA. http://betterthancash.org/wp-content/uploads/2013/12/Evidence-Paper-English1.pdf.

Bill and Melinda Gates Foundation. 2013. "Fighting Poverty, Profitably: Transforming the Economics of Payments to Build Sustainable, Inclusive Financial Systems." Special report, Financial Services for the Poor, Bill and Melinda Gates Foundation, Seattle, WA.

Bold, C., D. Porteous, and S. Rotman. 2012. "Social Cash Transfers and Financial Inclusion: Evidence from Four Countries." Focus Note 77, Consultative Group to Assist the Poor (CGAP), Washington, DC.

Brand South Africa. 2014. "South Africa's Social Grant Clean-Up Saves Billions." Brand South Africa website, February 7. http://www.southafrica.info/about/social/grant-070214.htm#.Vcwz1VL9Vus.

Camner, G. 2013. "Snapshot: Implementing Mobile Money Interoperability in Indonesia." Case study for the Mobile Money for the Unbanked Program, Groupe Speciale Mobile Association (GSMA), London.

CBA (Commercial Bank of Africa). 2013. "What Is M-Shwari." CBA web page: http://cbagroup.com/m-shwari/what-is-m-shwari/.

CBK (Central Bank of Kenya). 2016. *Banking Supervision Annual Report 2015*. Nairobi: CBK.

CGAP (Consultative Group to Assist the Poor). 2011. "CGAP G2P Research Project: Brazil Country Report." Research report, CGAP, Washington, DC.

Chevrollier, N., and S. Schmidt. 2014. "Overcoming the Last Mile Challenge: Distributing Value to Billions." *Forbes*, February 17.

Constantine, Giles. 2014. "Mexico Rebrands Flagship Social Welfare Programme in Bid to Help Working Poor." *Eye on Latin America* (blog), September 26.

Cook, T., and C. McKay. 2015. "How M-Shwari Works: The Story So Far." Forum 10, Consultative Group to Assist the Poor (CGAP), Washington, DC.

DailyFT. 2015. "Dialog Flies Lankan Flag High at Mobile World Congress with Two Global Awards." March 9.

Davidson, N., and P. Leishman. 2012. "Building a Network of Mobile Money Agents." Handbook for mobile network operators, Mobile Money for the Unbanked Program, Groupe Speciale Mobile Association (GSMA), London.

di Castri, S., and L. Gidvani. 2014a. "Enabling Mobile Money Policies in Tanzania: A 'Test and Learn' Approach to Enabling Market-Led Digital Financial Services." Case study, Groupe Speciale Mobile Association (GSMA), London.

———. 2014b. "The Kenyan Journey to Digital Financial Inclusion." Infographic, Groupe Speciale Mobile Association (GSMA), London. https://www.gsma.com/mobilefordevelopment/wp-content/uploads/2014/09/MMU_2014_Kenya-Pathway_Infographic_Web.pdf.

diGIT IT (magazine). 2014. "Sri Lanka Mobile Handsets Market Review." June 20.

Engel, M., and J. Scher. 2014 "Four Barriers—and Four Solutions—to Financial Inclusion through Payment Innovations." Center for Financial Inclusion (blog), December 15. http://cfi-blog.org/2014/12/15/four-barriers-and-four-solutions-to-financial-inclusion-through-payment-innovations/.

Ernst & Young. 2014. "Mobile Money—The Next Wave of Growth: Optimizing Operator Approaches in a Fast-Changing Landscape." Study report, Ernst & Young, London.

FII (Financial Inclusion Insights). 2015. "Indonesia: Survey of Users and Nonusers of Financial Services." InterMedia FII Wave Report, FII, Washington, DC. http://finclusion.org/uploads/file/reports/InterMedia-FII-Indonesia-wave-report-2-24-2015-updated.pdf.

FinScope. 2014. "FinScope South Africa, 2014—Survey Highlights." FinMark Trust, Johannesburg, South Africa.

Flaming, M., C. McKay, and M. Pickens. 2011. "Agent Management Toolkit: Building a Viable Network of Branchless Banking Agents." Technical guide, Consultative Group to Assist the Poor (CGAP), Washington, DC.

FSD (Financial Sector Deepening) Kenya and CBK (Central Bank of Kenya). 2013. "FinAccess National Survey 2013: Profiling Developments in Financial Access and Usage in Kenya." Survey results report, FSD Kenya and CBK, Nairobi.

Gathara, Victor. 2013. "Linda Jamii: A Partnership to Unlock the Value of eHealth." Presentation at AITEC East Africa Summit, Nairobi, November 20.

Gencer, Menekse. 2009. "Developing the Business Case for Your MMT Service." Presentation on behalf of mPay Connect to the Asia Pacific Mobile Money Transfer Conference, Manila, Philippines, December 7.

GSMA (Groupe Speciale Mobile Association). 2014. "State of the Industry 2014: Mobile Financial Services for the Unbanked." Report of the GSMA Mobile Money for the Unbanked (MMU) Program, GSMA, London.

IFC (International Finance Corporation). 2015. "Achieving Interoperability in Mobile Financial Services: Tanzania Case Study." IFC case study, World Bank Group, Washington, DC.

ITU (International Telecommunication Union). 2012. "Strategies for the Promotion of Broadband Services and Infrastructure: A Case Study on Sri Lanka." ITU and Broadband Commission for Digital Development, Geneva.

Jack, W., and T. Suri. 2014. "Risk Sharing and Transactions Costs: Evidence from Kenya's Mobile Money Revolution." *American Economic Review* 104 (1): 183–223.

Johnson, S. 2012. "The Search for Inclusion in Kenya's Financial Landscape: The Rift Revealed." Report commissioned by FSD Kenya, Nairobi.

Kumar, K., and M. Tarazi. 2012. "Interoperability in Branchless Banking and Mobile Money." CGAP (Consultative Group to Assist the Poor) blog, January 9. http://www.cgap.org/blog/interoperability-branchless-banking-and-mobile-money-0.

Maina, S. 2017. "Safaricom FY2017: Data and M-Pesa Were Safaricom's Biggest Earners." *Techweez*, May 10. http://www.techweez.com/2017/05/10/safaricom-fy-2017-data-m-pesa/.

Mas, I., and O. Morawczynski. 2009. "Designing Mobile Money Services: Lessons from M-PESA." *Innovations* 4 (2): 77–91.

Matinde, V. 2013. "Interview with Danson Muchemi, Founder of JamboPay." *Web Africa*, August 16. http://www.itwebafrica.com/home-page/movers-and-shakers/548-danson-muchemi/231484-interview-with-danson-muchemi-founder-of-jambopay.

Miller, P., and D. Salazar. 2013. "Expanding Card Acceptance to Small Merchants Globally through Mobile Point of Sale (MPOS)." White paper, MasterCard Advisors, Purchase, NY.

Morawczynski, O. 2015. "Just How Open Is Safaricom's Open API?" CGAP (Consultative Group to Assist the Poor) blog, http://www.cgap.org/blog/just-how-open-safaricom's-open-api.

Mostafa, J. 2014. "Cadastro Único: A Registry Supported by a National Public Bank." IPC One Pager No. 250, International Policy Center for Inclusive Growth, United Nations Development Programme, Brazil.

National Treasury, South Africa. 2013. "Social Security and the Social Wage." In National Budget 2013 Review, National Treasury, Republic of South Africa, Pretoria.

Omwansa, T., and N. Sullivan. 2012. *Money, Real Quick: The Story of M-PESA*. Norwich, UK: Guardian Shorts.

Paes-Sousa, R., L. Santos, and E. Miazaki. 2011. "Effects of a Conditional Cash Transfer Programme on Child Nutrition in Brazil." *Bulletin of the World Health Organization* 89 (7): 496–503.

Pénicaud, C., and A. Katakam. 2014. "State of the Industry 2013: Mobile Financial Services for the Unbanked." Report of the GSMA Mobile Money for the Unbanked (MMU) Program, GSMA, London.

RBA (Retirement Benefits Authority). 2013. "Mbao Pension Plan." RBA web page: http://www.rba.go.ke/index.php/en/component/content/article?id=55.

Safaricom. 2013. "H1 FY14 Presentation." PowerPoint presentation, Nairobi, November 5. https://www.safaricom.co.ke/images/Downloads/Resources_Downloads/Half_Year_2013-2014_Results_Presentation.pdf.

———. 2017. "Celebrating 10 Years of Changing Lives." Interactive M-Pesa milestones timeline, https://www.safaricom.co.ke/mpesa_timeline/timeline.html.

Samson, M., K. Mac Quene, and I. van Niekerk. 2006. *Designing and Implementing Social Transfer Programmes*. Cape Town: Economic Policy Research Institute.

SASSA (South African Social Security Agency). 2013. "Annual Report 2012/2013." http://www.sassa.gov.za/index.php/knowledge-centre/annual-reports.

Schonhart, S. 2015. "Mobile Banking Struggles to Gain Traction in Indonesia." *Wall Street Journal* online, July 21. http://www.wsj.com/articles/mobile-banking-struggles-to-gain-traction-in-indonesia-1437507127.

Syngenta Foundation. 2011. "Foundations and Partners Extend Insurance Reach." *Syngenta Foundation News*, February 25, Syngenta Foundation for Sustainable Agriculture, Basel, Switzerland.

UN (United Nations). 2014. *United Nations E-government Survey 2014: E-government for the Future We Want*. New York: UN.

Volker, W. 2013. *Essential Guide to Payments: An Overview of the Services, Regulation and Inner Workings of the South African National Payment System*. Pretoria: Veritas Books.

Wetzel, D. 2013. "Bolsa Família: Brazil's Quiet Revolution." News item, World Bank website: http://www.worldbank.org/en/news/opinion/2013/11/04/bolsa-familia-Brazil-quiet-revolution.

World Bank. 2012a. "Innovations in Retail Payments Worldwide: A Snapshot. Outcomes of the Global Survey on Innovations in Retail Payment Instruments and Methods." Consultative Report, World Bank, Washington, DC.

———. 2012b. "Developing a Comprehensive National Retail Payments Strategy." Consultative Report, World Bank, Washington, DC.

———. 2014. "A Model from Mexico for the World." Feature story, World Bank website, November 19. http://www.worldbank.org/en/news/feature/2014/11/19/un-modelo-de-mexico-para-el-mundo.

CHAPTER 6

Unique Identification

Introduction

Unique national identification (ID) is a critical component of a digital money system. The ability of the service provider to accurately identify its current and potential customers is central to providing digital-money services, such as mobile money or prepaid-card-based government grant programs. One important reason why M-Pesa was able to roll out the mobile money initiative so quickly—while also addressing the know-your-customer (KYC) concerns—is because Kenya has an established national ID system.

In Sri Lanka too, mobile network operators (MNOs) leveraged the existing national ID system put in place by the national government and the mandatory registration of subscriber identification module (SIM) cards when launching mobile money solutions. The Central Bank of Sri Lanka relaxed its KYC requirements, adopting a more proportionate approach to customer due diligence (CDD) based on the e-wallet size. Similarly, Thailand's biometric smart ID enabled wider reach in banking access, while in South Africa, biometrics were introduced to the social security system, to register or identify grant recipients.

The success of mobile money and scaling-up depends on speeding up account opening and minimizing steps in transacting. Overly strict requirements regarding the identification and verification of customers tend to restrain the impact of efforts to increase financial inclusion. For example, strict application of CDD requirements may exclude people lacking official documentation from entering the regulated financial system. Strict CDD procedures may also lead financial institutions to pass on costs to the customer, resulting in a disincentive (especially for the poor) and thus may push them toward informal service providers. Although such measures may be appropriate risk mitigation measures for regulated financial institutions, they may result in marginalization and exclusion of poor people from financial intermediary functions and payments mechanisms that enhance poor peoples' coping capability.

In terms of Financial Action Task Force (FATF) guidelines as endorsed by the Bank for International Settlements (BIS), from an Anti-Money Laundering and

Combatting Funding of Terrorism (AML/CFT) perspective, small-value and low-risk mobile money transactions qualify for simplified CDD. Therefore, countries may consider applying "tiered" approaches whereby KYC and CDD requirements vary according to transaction or payment limits or wallet sizes. Higher limits entail a more extensive CDD process, including proactive account monitoring for suspicious activities. Since most potential mobile money users or grant recipients are either unbanked or underbanked, it is important to maintain minimum KYC requirements and allow simplified CDD processes to sign up for mobile money and digitally maintained grant programs, and to transfer or transact smaller values. At the entry level, a national ID document can be used to validate identity for KYC purposes. Sri Lankan mobile money operators use this tiered KYC/CDD approach and offer several types of accounts with different KYC requirements to ease registration requirements for low-income users.

In developing countries, civil registration systems are largely absent or cover only a fraction of the population. In developed countries, most people establish their official identity through an official birth certificate. However, a United Nations Children's Fund (UNICEF 2013) report shows that births of nearly 230 million children under five have never been registered—approximately one in three of all children under five around the world. UNICEF estimates that 35 percent of children worldwide, and 40 percent of children in developing countries, were not registered at birth. South Asia had the highest percentage of unregistered births (61 percent) in 2013, followed by Sub-Saharan Africa (56 percent) (Gelb and Clark 2013).

UNICEF maintains that having a birth registration is the first step toward realization of other rights throughout a person's lifetime. Furthermore, the UN High Commissioner for Refugees (UNHCR) estimates that at least 10 million people across all regions are stateless (UNHCR 2017). Without official ID, these people formally do not exist. Gelb and Clark (2013) explain that this "identity gap" is increasingly recognized not only as a symptom of underdevelopment but also as a factor that makes development more difficult and less inclusive. They suggest that biometric ID systems could have a transformative impact on development, similar to microcredit and mobile phones in the far-reaching ability to improve poor people's lives.

Biometrics take a variety of forms—iris scans, fingerprints, even face scans—and are being used to establish identities for many different purposes: elections, health care, payroll, government services, and more. Gelb and Clark (2013) conducted the first global survey of 160 cases where biometric ID has been used for economic, political, and social purposes in developing countries (map 6.1). The survey results show that some 1 billion people in developing countries across the world have taken part in biometric ID programs. Sub-Saharan Africa is the most active region (75 surveyed cases, covering an estimated 288 million people), while South Asia is third (27 surveyed cases, covering an estimated 426 million).[1] Overall, at least half of these projects are donor-funded with official development assistance.

Map 6.1 Global Participation in Biometric ID Programs, by Region, 2012

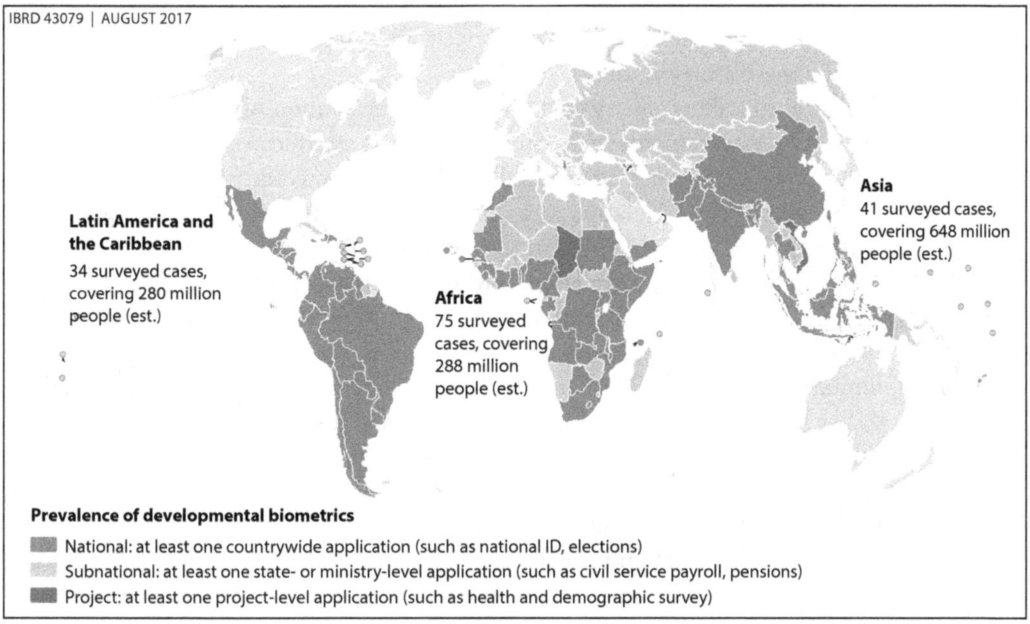

Source: Gelb and Clark 2013. © Center for Global Development (CGD). Reproduced, with permission, from CGD; further permission required for reuse.

The cases are categorized as either (a) ID-supply-driven, or foundational, cases (such as establishing civil registries and national IDs); or (b) application-demand-driven, or functional cases (such as demand for particular services or transactions such as voter IDs, bank cards, and health records). Under the functional class, the general financial services category is an additional transactional type needing ID. In terms of projects or businesses that are implementing ID-facilitated financial services, Sub-Saharan Africa is in the lead with eight projects, followed by South Asia with three projects. Biometric authentication appears not to be widely used in mobile banking, although certain smartphones already have fingerprint scanners.[2] While currently biometrics have not been popular in the developing-country mobile applications, they are available as a tool to expand use over time.

Biometrics are used more extensively in a wide variety of social transfers. The survey identifies 23 cases worldwide where biometrics have been used in creating beneficiary registries and authenticating cash or in-kind transfers at the point of service.

Biometric ID technology is one option for identity management and identification that can help to leapfrog traditional paper-based identity systems in terms of proliferation, efficiency, and accuracy. It can extend current security and anti-fraud efforts by preventing ready impersonation of the authorized user. At the same time, practical difficulties in implementing biometric ID cannot be ruled out, including the huge up-front cost of registering the users, privacy issues, and social and cultural issues involved in getting people to consent to biometric ID methods. A further question is how well the data security and safeguard measures, and other operational aspects in using biometrics, would work in developing-country settings.

In the mobile money world, mandatory registration of prepaid SIM cards is viewed by governments and regulators as a counterterrorism measure enabling efficient KYC assessment, especially in developing countries. In the developed world where self-regulation is highly valued, privacy rights have discouraged such practices. As highlighted in the Groupe Speciale Mobile Association white paper (GSMA 2013), mandating SIM registration can have unintended negative consequences, such as loss of access to communications services when mobile users' SIM cards are deactivated; restricted access due to possible geographic limitations on where SIM cards can be registered or purchased; and emergence of black markets for fraudulently registered or stolen SIM cards. However, establishing such a registry opens up more options for the users. Hence, it is up to the policy makers to harmonize the regulations with market needs after consulting with industry to analyze costs, benefits, and implementation options.

A national ID system provides not only verification of a person's identity, but also addresses acceptable and approved basic KYC protocols for accessing finance and payment systems. The case studies analyzed in this chapter highlight why having a unique ID (UID) is a foundational and critically important enabler toward scaling up e-money initiatives while managing risks. The case studies focus on

- How lack of a UID can be a limitation in terms of scaling up digital finance for disaster relief programs and financial inclusion schemes;
- How transformative responses to the ID barrier enable millions of people to access financial services in a hassle-free manner; and
- How digital identity enhances customer experience in moving toward a cash-lite society.

Not directly addressed in this volume are the political economy dimensions where vested interests actively hinder development, application, spread, and use of a unique identifier. By definition, a UID in any environment allows traceability by the tax authorities; enables AML/CFT compliance; and eliminates opportunities for leakages (fraudulent receipt of subsidies and social transfers), black market transactions, tax avoidance, and theft. Linkage of UID across financial sector databases and with phone and utility authorities promotes transparency and traceability, which are also challenges.

The Philippines: 21 IDs and Counting

If unbanked individuals lack appropriate forms of identification, which is not uncommon, they may struggle to participate in mobile money opportunities. This is prompting many governments to rethink the identity process and use creative solutions that employ biometrics, for example, or that use group gatherings, where the identified identify the unidentified.

—USAID-Citi Mobile Money Accelerator Alliance,
"10 Ways to Accelerate Mobile Money"

There is no unique national ID system in the Philippines. Various government departments collect personal information about every Filipino who has been issued a passport, a driver's license, or a residence certificate. The government and businesses also obtain personal information for income tax payments, voter registration, marriage, public health insurance coverage, security clearance for job applications, and most business transactions.

The government initiated actions to establish a national ID system two decades ago, but various interest groups have thwarted implementation. During her term President Gloria Macapagal Arroyo issued an executive order for government agencies and government-controlled corporations to harmonize their systems, and the unified multipurpose ID card (UMID) was established. However, this excluded nongovernment workers and average citizens, including those who are unemployed or working abroad. Proposals to issue every citizen a national ID card were vehemently opposed, principally on the perceived threat to security and violation of privacy by the government's collection of personal information. At the same time, proponents argued that this could help fight criminal activities and cut down on corruption.

The Central Bank of the Philippines (Bangko Sentral ng Pilipinas, or BSP) has issued a circular identifying 21 different types of valid ID that can be used for financial transactions,[3] and several other ID types are also accepted in the Philippines (table 6.1). The length of the list demonstrates how difficult the ID process would be for the customer and service provider, as well as for the regulator, to provide and verify one or more of these forms of ID, whose authenticity is often difficult to establish.

Anecdotal evidence suggests that providing a valid ID is still difficult for most of the poor, who tend to prefer informal financial service providers that use group or personal identification methods rather than subjecting themselves to KYC checks by regulated formal financial institutions. In rural areas, 558 rural and cooperative banks with a network of 2,101 branches provide basic intermediary functions. Poor people in the Philippines often use sari-sari shops[4] or pawnshops for their financing needs. Many (25 percent) still use drivers and relatives for money transfer purposes.

Sari-sari stores are ubiquitous in the Philippines, numbering over a million, nearly all of them owned by women. A demand study on domestic payment systems done by Bankable Frontier Associates (BFA 2010) revealed that pawnshops (such as LBC, ML, Cebuana Lh., and others) are well known, and more people use them for their payment needs than other service providers (figure 6.1). In 2014, there were 5,971 pawnshops with a branch network of 11,542 shops in the country (BSP 2014). Even though

Table 6.1 Acceptable ID Documentation for Financial Services in the Philippines

	Form of ID	*Sample*
Listed in BSP Circular No. 608 Series of 2008		
1	Passport	
2	Driver's License	
3	Professional Regulation Commission ID	
4	National Bureau of Investigation Clearance	
5	Police Clearance	
6	Postal ID	
7	Voter's ID	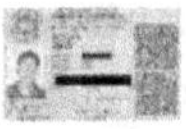
8	Barangay Clearance/Certificate	
9	Government Service Insurance System ID	
10	Social Security System ID	
11	Senior Citizen's ID Card	
12	Overseas Worker's Welfare Administration ID	
13	Overseas Filipino Worker ID	
14	Seaman's Book	

table continues next page

Table 6.1 Acceptable ID Documentation for Financial Services in the Philippines *(continued)*

	Form of ID	*Sample*
15	Alien Certificate of Registration	
16	Government Office or Government Owned and Controlled Corporations ID (such as AFP ID, HDMF [Pag-IBIG Fund] ID, and others)	
17	Certification from the National Council for the Welfare of Disabled Persons	—
18	Department of Social Welfare and Development ID	
19	Integrated Bar of the Philippines ID	
20	Company ID	—
21	Student's ID or School ID (as beneficiary for remittances or fund transfers)	—
Other types of valid IDs not listed in the 2008 BSP circular		
1	Tax Identification Number (TIN ID)	
2	National Statistics Office (NSO) Birth Certificate	
3	Marriage Certificate (NSO Authenticated)	
4	PhilHealth Identification Card	
5	Consular ID	—
6	Permit to Carry Firearms	
7	Company/Office ID	—
8	Philippine Overseas Employment Association ID	—
9	PRA Special Resident Retiree Visa ID	—
10	Unified Multipurpose ID (UMID)	—

Source: "Top 31 Valid ID's Required in the Philippines," affordableCebu website: http://www.affordablecebu.com/load/philippine_government/top_28_valid_id_39_s_required_in_the_philippines/5-1-0-109#ixzz3bY8UUSZJ.
Note: BSP = Central Bank of the Philippines. — = not available.

BSP was given the authority and responsibility to regulate pawnshops since 1972, they remain the least-regulated businesses that fall under BSP.

For service providers, too, lack of a universally accepted ID is a barrier when it comes to account opening and operating. With no universal national ID, the financial sector must rely on other forms of ID, which not all customers may have. The institutions have to manage their risk profiles while adhering to KYC

Figure 6.1 Use and Awareness of Payment System Providers in the Philippines, 2010

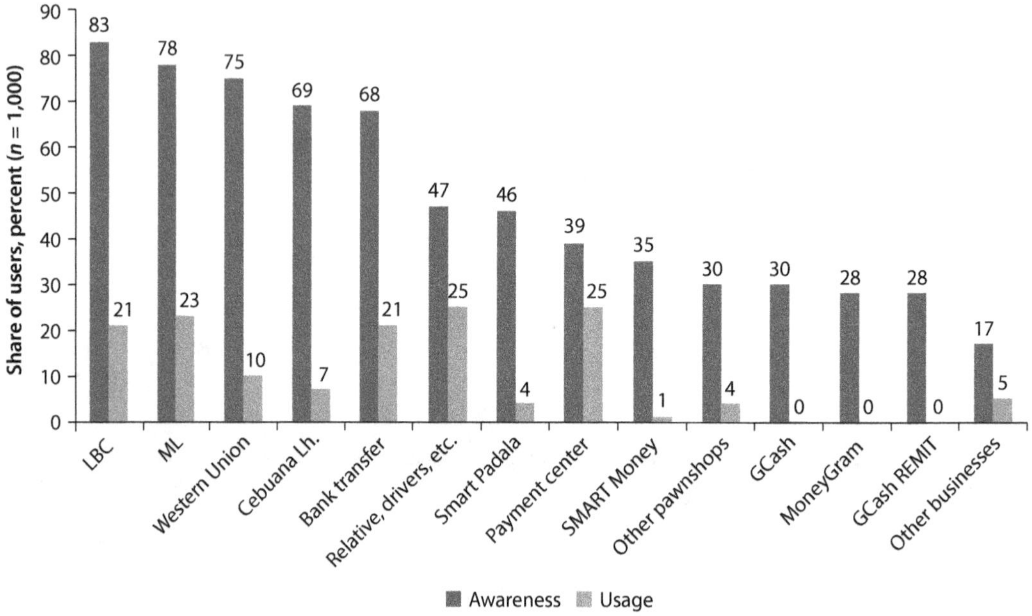

Source: BFA 2010.
Note: LBC Express is the market leader in the Philippines for payments, remittances, courier services, and so on. ML = M Lhuillier Bank.

and CDD guidelines set out by the authorities. Additional training and capacity building will be needed to carry out an acceptable level of due diligence. Regulators can set risk-based measures such as limited transaction amounts and reporting thresholds to ensure financial access. Also, the agents will have to be given clear guidelines and training on performing KYC and CDD duties using acceptable ID documents. Again, anecdotal evidence shows that confusion remains widespread, making customers apprehensive about going to a formal financial institution.

UID is important in the provision of government welfare disbursements and disaster recovery management. Using a UID to verify identification and entitlement, and making the payments accordingly, could address the leakage and inefficiencies in some transfer programs based on cash. Coupled with switching to electronic payments, this could save the government a considerable amount of money that could be channeled back to the welfare schemes. The Philippines is one of the most disaster-prone countries in the world. Hence, streamlining its disaster recovery program in terms of providing assistance is important for the country. In the aftermath of super typhoon Haiyan in 2013, the Philippines' primary statistics office revived calls for a national ID system to speed up the process of identifying calamity victims—both the dead and the missing—and to offer financial assistance to the surviving victims.

The Philippines had begun to move a step closer to getting a UID card. The House endorsed Bill No. 5060 (the proposed Filipino Identification Act) for Senate passage on May 20, 2015, which had been expected to greatly streamline government transactions and promote more efficient service delivery. The proposed ID system would bring all existing government-initiated ID systems into a single integrated ID system. Once effectively implemented, the ID will provide proof of identity, status, age, and address for admission to all learning institutions, employment purposes, voting identification, transactions in banking and financial institutions, and provision of benefits or privileges afforded by law to senior citizens.

The House of Representatives had also approved a bill requiring the registration of all prepaid mobile-phone SIM cards. If Bill 5231 had become law, it would cover tens of millions of existing and future prepaid mobile-phone users. At present, anyone can buy a prepaid SIM card and use it without filling out any form or being asked to present an identification card. Under the bill, as proposed, sale may be denied if the buyer does not present valid and clear proof of identity. In most countries SIM card registration with national ID is taken as the basic threshold in setting risk-based tiered systems for KYC and CDD purposes. As of May 2017, the Filipino Identification System (FilSys) had not yet been passed into law.

By beginning to seriously consider the UID issue, the Philippines is addressing one of the key conditions for mobile money proliferation. In a country already facing unique geographical barriers to financial access, enabling government programs to go digital and mobile offers a crucial step toward financial inclusion. Steered by BSP as a forward-thinking and enabling regulator, the effort to lower these macro-level barriers should improve the ability of the unbanked population to access the mainstream financial sector in pursuit of economic opportunities.

India's Aadhaar Program: Potential Game Changer in Digital Financial Inclusion

The largest and best-known biometric ID program in the world is India's UID program, which provides a unique 12-digit identity number called "Aadhaar" to 855 million people (around 68 percent of the country's total population) as of May 2015 (figure 6.2).

In the year 2015 alone, 124 million people were admitted to the program at a rate of about 24 million new identifications each month. Uttar Pradesh has the highest number of registrations (118 million), followed by Maharashtra (93.5 million), and West Bengal (63 million) (figure 6.3).

Genderwise, out of the total number of Aadhaar registrations, 51.8 percent are male, while female coverage is 48.2 percent. Out of the total, 77.3 percent are age 18 and above (figure 6.4). These numbers are truly impressive and remarkable, especially considering that signing up for Aadhaar ID is not mandatory. Aadhaar can affect financial inclusion and inclusive growth in a transformative manner by establishing links with bank accounts and facilitating a variety of services targeted at the poor.

Since issuing the first Aadhaar number on September 29, 2010, in Maharashtra state, the Unique Identification Authority of India (UIDAI) remained on track

Figure 6.2 Aadhaar Registration Trends in India, 2014–15

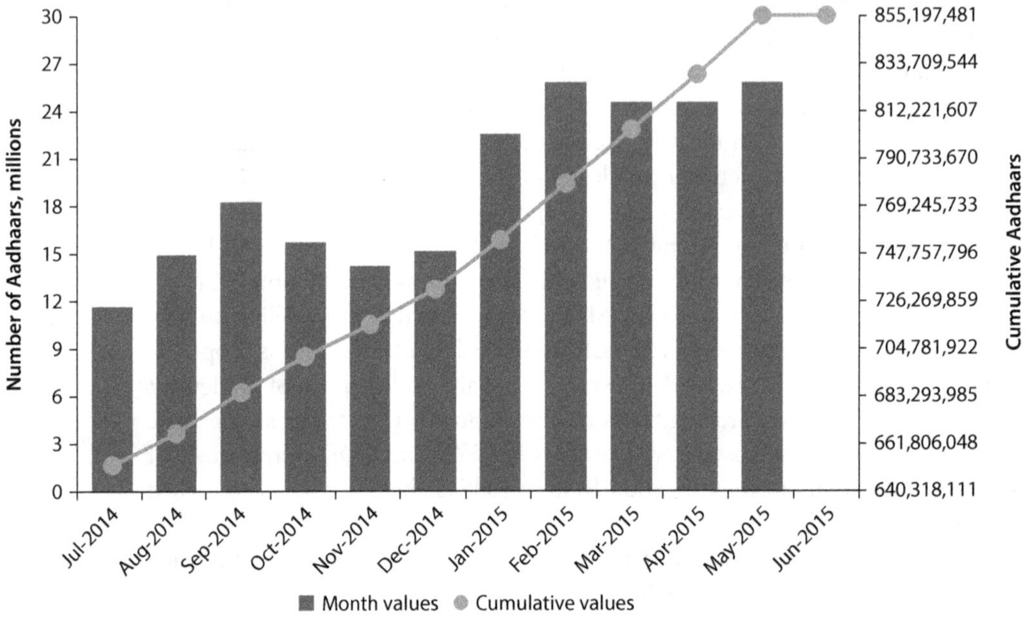

Source: "Aadhaar Dashboard," Unique Identification Authority of India (UIDAI), https://portal.uidai.gov.in/uidwebportal/dashboard.do.

Figure 6.3 Top 10 States for Aadhaar Registration in India, 2015

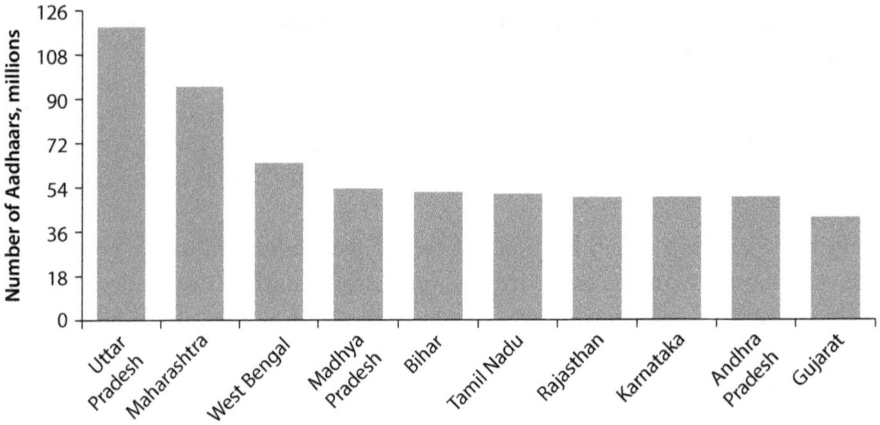

Source: "Aadhaar Dashboard," Unique Identification Authority of India (UIDAI), https://portal.uidai.gov.in/uidwebportal /dashboard.do.

Figure 6.4 Aadhaar Registration, by Gender and Age Group in India, 2015

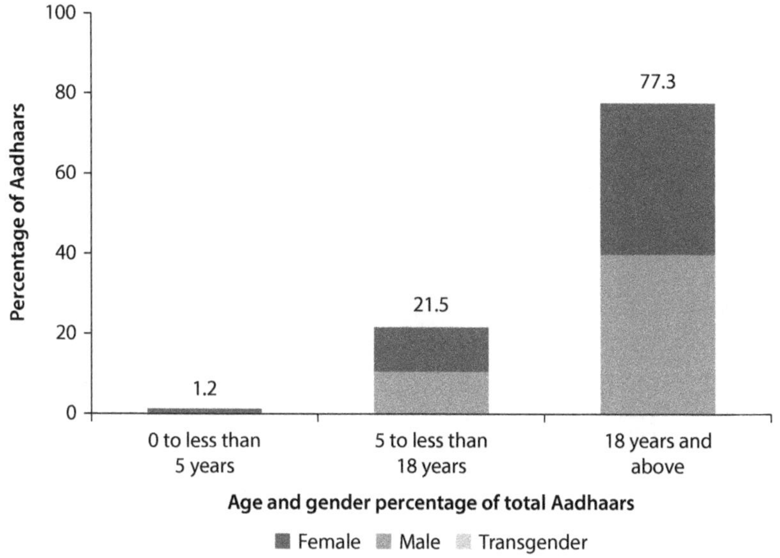

Source: "Aadhaar Dashboard," Unique Identification Authority of India (UIDAI), https://portal.uidai.gov.in/uidwebportal/dashboard.do.

to achieve its target of issuing Aadhaar numbers to 930 million individuals by the end of 2015. To reach such a massive scale, UIDAI used over 275 service providers chosen through a competitive process in line with procurement norms laid down by the government (figure 6.5). Both Indian and non-Indian organizations have contributed significantly to the proliferation of Aadhaar. In addition, by December 2011, UIDAI had true and tested statistics demonstrating system accuracy of 99.965 percent computed from an operational resident enrollment database of 84 million people (UIDAI 2012b).

UIDAI was established under the Planning Commission of India with a budget of Rs 123.98 billion (US$1.94 billion). By July 2013, Rs 30.62 billion had been spent on the project (*Economic Times* 2013). The National Institute of Public Finance and Policy conducted a cost-benefit analysis of the Aadhaar project over a 10-year period (2010–21) and concluded that the Aadhaar project has an internal rate of return in real terms of 52.9 percent to the government (NIPFP 2012).

Aadhaar offers a platform approach to financial services, especially financial inclusion (figure 6.6), by providing standardized services such as

- Electronic KYC platform for Aadhaar-enabled accounts;
- A payments mechanism to facilitate direct benefits transfer (DBT) by government departments; and
- Standardized consumer experience through the Aadhaar Enabled Payment System (AEPS) and micro automated teller machines (ATMs).

Figure 6.5 Number of Aadhaar Registrations Completed by Top 10 Service Providers in India, May 2015

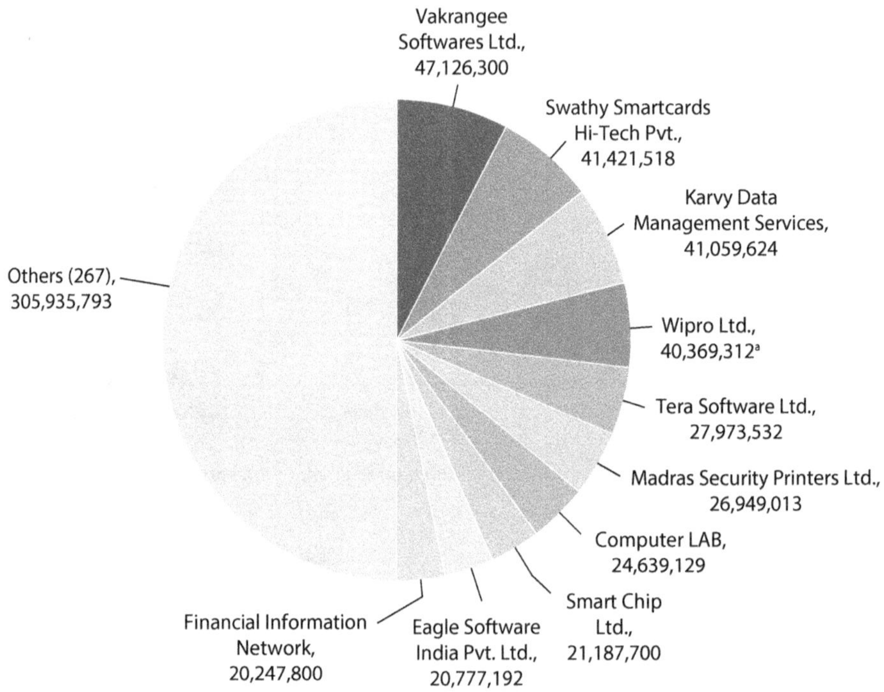

Source: "Aadhaar Dashboard," Unique Identification Authority of India (UIDAI), https://portal.uidai.gov.in/uidwebportal/dashboard.do.
a. The number of Wipro registrations is as of 2017.

Figure 6.6 Financial Inclusion Applications of Aadhaar

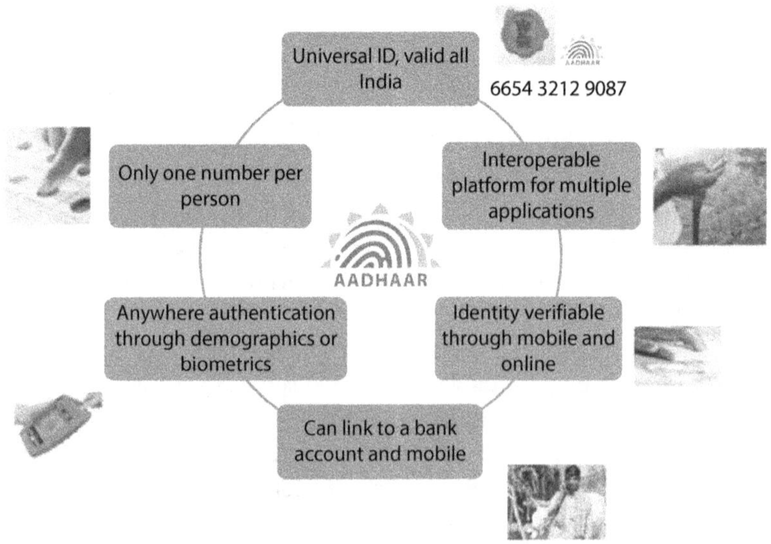

Source: UIDAI 2012a. © Unique Identification Authority of India (UIDAI). Reproduced, with permission, from UIDAI; further permission required for reuse.

AEPS allows Aadhaar cardholders to carry out financial transactions on a micro-ATM terminal (point of sale [POS]) at a business correspondent (BC).[5] The Aadhaar number, along with an individual's mobile number and an electronic bank account number, can revolutionize the way people access and transact financial products and services. Aadhaar has the potential to become the universal linkage for disbursing government payments in a more secure, streamlined, efficient, and cost-effective way.

e-KYC and Aadhaar-Enabled Accounts

With Aadhaar, the banks can now authenticate customer details with the UIDAI. The UIDAI provides an e-KYC service through which the KYC process can be performed electronically with explicit authorization by the cardholder. As part of the e-KYC process, the cardholder authorizes UIDAI (through Aadhaar authentication using either biometrics or a one-time password) to provide their demographic data along with their photograph (digitally signed and encrypted) to service providers such as banks (UIDAI 2016). This process facilitates easy and quick account opening at banks.

The Aadhaar-linked e-KYC also provides a hassle-free approach for the millions of poor in India who do not have other ownership or identity documents. The Reserve Bank of India (RBI) in September 2013 approved the paperless electronic authentication (e-KYC) provided by the UIDAI as valid ID documentation. Previously, in December 2012, the RBI allowed the use of the Aadhaar enrollment letter as a valid proof of identity and address (Rao 2013). These are important regulatory measures taken by the regulator to further financial inclusion.

Electronic opening of accounts at the time of Aadhaar enrollment is possible in partnership with banks through an electronic process. Aadhaar can also be linked to an existing bank account through any of the delivery channels—for example, branch, ATMs, Internet, mobile, and micro-ATMs. Any bank interested in using the facility must use UIDAI-approved biometric scanning devices and register with it to receive access to its data.

Direct Benefits Transfer

To facilitate DBT, bank accounts of the recipients can be linked with the Aadhaar number. This ensures that money under various developmental welfare schemes and subsidies reaches beneficiaries directly and without delay. The scheme was launched in the country in January 2013 and has been rolled out in a phased manner, starting with 26 welfare schemes in 43 districts. This was extended to an additional 78 districts and three more schemes from July 1, 2013, and is to be extended to the entire country in a phased manner (Srikanth 2013).

The government also started transfer of a cash subsidy for domestic liquefied petroleum gas (LPG) cylinders[6] (used for energy and fuel purposes by households) to Aadhaar-linked bank accounts of the customers from June 1, 2013, in 20 pilot districts. About 7.5 million recipients benefited in these districts.

A modified scheme was relaunched in November 2014 in 54 districts covering 23 million households, and will be phased into the balance of 622 districts in the rest of the country.

The scheme aims at halting diversion to other purposes of highly-subsidized LPG cylinders meant for domestic use. Over 153 million consumers across 676 districts of the country will be covered. Currently over 65 million consumers have already joined the scheme and will continue to receive the subsidy through their bank accounts. Since the modified program was launched in November 2014, an amount of Rs 6.24 billion has been transferred to over 2 million LPG consumers (MPNG 2014).

The Mahatma Gandhi National Rural Employment Guarantee Act (MGNREGA) is one of the flagship schemes of the Ministry of Rural Development (MoRD), which guarantees 100 days of wage employment in a financial year to every rural household whose adult members volunteer to do unskilled manual work. Since September 2013, MoRD has been actively working toward collecting Aadhaar data of its beneficiaries, entering the data into NREGASoft (software used by MGNREGA) and beneficiary bank accounts, and thereafter making payments through the Aadhaar Payment Bridge System. So far, providing the Aadhaar ID is optional. By October 2014, 20.7 million out of 105 million beneficiaries of MGNREGA had already linked their Aadhaar accounts to the program (MGNREGA 2014). According to news articles, Rs 150 billion will be disbursed annually to around 43 million MGNREGA beneficiaries (Mehra 2015).

Using the DBT facility, all types of social welfare schemes offered by different government departments can be streamlined and managed through a single Aadhaar-linked account, thus creating efficiency gains for the government while reducing opportunity and accessibility cost to the recipients. An International Monetary Fund report (IMF 2013) states that the total savings from linking DBT to Aadhaar bank accounts could be substantial. The report highlights that leakages from outdated biographical information, "ghost" identifications, double registration, and other losses are estimated in the range of 15–20 percent of total spending. If the combination of direct cash transfer and Aadhaar eliminates the estimated 15 percent leakage cited above for the programs being integrated, savings could total 0.5 percent of gross domestic product (GDP), in addition to the gains from the better targeting of spending on the poor.

Micropayments and the Aadhaar-Enabled Payment System

To enable payments from the Aadhaar-linked accounts, banks are issuing debit cards to beneficiaries. Banks have also started strengthening banking infrastructure, providing BCs in areas that were previously unserved, and introducing mobile banking solutions to ensure access and facilitate payments over a wide geographic area. The branch networks with ATMs, along with mobile banking systems and BC agents, ensure that banking services reach all the beneficiaries.

The cost of providing banking services to the poor, who transact in small amounts, is a limitation. Banks consider such "micropayments" unattractive because the transaction costs are relatively high. Aadhaar enables an efficient, cost-effective payment solution (AEPS) for promoting financial inclusion.

AEPS is a bank-led model that allows online interoperable financial inclusion transactions at POS (also called micro-ATMs) that are operated by a BC or BC subagent, appointed by banks using the Aadhaar authentication. Unlike an ATM, the cash-in/cash-out functions of the micro-ATM are performed by an operator, thus bringing down the cost of the device and the cost of servicing the customer. The micro-ATM supports deposits, withdrawals, fund transfers, and balance inquiries. Once successfully deployed, the AEPS payment mechanism will allow account access from anywhere through any delivery channel online in real time, thus enabling an efficient and cost-effective remittance transfer system within India. In addition, a variety of other services such as microcredit, micro-insurance, and micropensions can be offered on top of this solution.

Though Aadhaar was a policy initiative by the previous government, it has received support from the present government. The Prime Minister's Office (PMO) has directed the Planning Commission to collect data on Aadhaar and DBT schemes in 300 priority districts where Aadhaar enrollment is currently over 60–70 percent, as well as information on Aadhaar and DBT with respect to five key government schemes: MGNREGA, pensions, scholarships, Public Distribution System, and LPG (Tewari 2014). The Indian government subsidizes food, fuel, and fertilizers, with the total subsidy bill earmarked in the 2015 budget at a massive Rs 2.5 trillion (around 2 percent of GDP). A significant portion of these funds is believed to be wasted because of improper targeting, middlemen, corruption, and duplication. Hence the Aadhaar-enabled DBT system could eventually address this massive leakage.

Issues and Setbacks with Aadhaar

Aadhaar is not without its critics among privacy and security advocates and others. Privacy and information security are very real concerns in a civil society, hence the need for open dialogue and providing assurance, especially given the way e-money in India is poised to take off with the introduction of Jan Dhan Yojana (JDY) accounts linked to the Aadhaar ID (as discussed further in chapter 4).

To address such concerns, a simplified form of security policy and standards should be introduced at inception so that service providers as well as users understand the potential risks in digital finance and adhere to security guidelines and procedures. Given that the system is dedicated to the poor and underserved groups, a disruption could have far-reaching political, economic, and social consequences. Other issues and concerns about Aadhaar are discussed in the next two subsections.

Aadhaar and the National Population Register

In addition to UIDAI and Aadhaar, the government of India initiated the National Population Register (NPR) in 2010, entailing the creation of a

Table 6.2 Key Differences between Aadhaar and the National Population Register

Aadhaar	*National Population Register*
Voluntary	Mandatory
UIDAI issues a number	Only a register
For identity and authentication	Signifies resident status and citizenship
Based on other forms of documentation and identification	Based on census information

Note: UIDAI = Unique Identification Authority of India.

National Citizens Register, which is being prepared at the local, subdistrict, district, state, and national levels. The database will contain 13 categories of demographic information and 3 categories of biometric data collected from all residents ages five years and above. Collection of this information was initially supposed to take place during the house listing and housing census phase of Census 2011 from April to September 2010 (Hickok 2013). The NPR is legally enshrined in the provisions of the Citizenship Act 1955 and the Citizenship Rules 2003.

While it is not clear why two separate exercises are needed, the Planning Commission is of the view that it is necessary to speed up the collection of biometric details of residents and issue them Aadhaar numbers as well as National Multi-purpose Identity Cards. The key differences between the two programs are shown in table 6.2.

Cost of Making Existing Infrastructure Aadhaar-Enabled

While the innovations have helped in generating the world's cheapest ID cards under the UIDAI—costing around Rs 100 per person (*Economic Times* 2014)—the challenge will be deployment of Aadhaar-compliant biometric fingerprint POS devices (micro-ATMs) at hundreds of thousands of agents. This can be a costly and complex proposition, making it challenging for the system to be scaled up quickly in a cost-effective manner.

In addition, given the financial inclusion mandate, although the government wants biometric authentication for credit card and ATM transactions, the market players are hesitant because of the costs they have to incur to upgrade each and every ATM and POS at thousands of merchant outlets. The RBI-constituted panel has estimated the cost of banks' readiness for Aadhaar at around Rs 42.6 billion. It is interesting to note that, other than the cost, the stakeholders agree on most things, and one cannot rule out the potential internal demand that is being created by the government through this massive financial inclusion project.

The Indian economy predominantly uses cash, with cash in circulation in 2013 at around 13 percent of GDP (more than double the global average) (Chakrabarty 2014). Studies have estimated that the cost of cash to the economy is significant (around 5–7 percent of GDP). This cost could be reduced by digitizing payments.

Aadhaar has come a long way from its 2009 inception and is on track to deliver transformational outcomes to India's poor. Financial inclusion is one such application riding on the UIDAI's infrastructure, since possession of an Aadhaar number enables opening of a bank account. Unlike a cash-in/cash-out e-wallet in an MNO-led mobile money solution, hundreds of millions of potential micro-entrepreneurs who were outside the banking system are being brought into the credit economy.

Some critics maintain that UIDAI has no legal standing. However, the present government is placing considerable emphasis on its financial inclusion agenda and plans to accelerate DBT payments through JDY bank accounts linked to the Aadhaar number. Hence it is likely that the government may resolve this issue, perhaps through a legislative enactment conferring statutory authority to UIDAI.

Sri Lanka: Mobile Connect, the Interoperable ID

> *Mobile Connect is your new online identity. Simply by matching you to your mobile phone, Mobile Connect lets you log-in to websites and applications quickly without the need to remember passwords and usernames. It's safe, secure and there's no need to worry that your personal information will be shared without your permission.*
>
> —Groupe Speciale Mobile Association (GSMA), "Introducing Mobile Connect"

The importance of a UID for success in scaling up mobile money or digitized government benefit programs was discussed in the two case studies earlier in this chapter. Although in the real world, a person's identity can be established and verified using identity cards and documents along with personal face-to-face verifications, in the digital world identity verification has to rely on digital identities. A digital identity is an online or networked identity adopted or claimed in cyberspace by an individual, organization, or electronic device. These users may also project more than one digital identity through multiple communities. In terms of digital identity management, key areas of concern are security and privacy.

Because digital identities created by individuals are inherently weak, and such identities are linked to online websites, e-mail addresses, or domains, security becomes a key consideration. The Mobile Connect Sri Lanka case study focused on how digital identity management allows consumers to move seamlessly from cash to cash-lite transactions through digital services. This is neither a costly operation nor meant only for expensive transactions. On the contrary, for average consumers, digital identity solutions such as Mobile Connect offer privacy protection, reduce the risk of identity theft, and simplify the login experience for a range of services, such as retail, health care, government, and banking, among others (Finextra 2014).

The Mobile Service Providers

In Sri Lanka, mobile penetration has grown steadily from 87 percent in 2011 to 123 percent in 2016 (Harpur 2016). Sri Lanka also has the world's lowest rates for fixed broadband Internet, around US$5 per month, resulting from a high level of competition among the suppliers and the government's explicit aim to keep the cost to a minimum (ITU 2012).

There are currently five mobile operators in Sri Lanka. Dialog Axiata, the market leader, and Sri Lanka Telecom Mobitel have a combined market share of over 65 percent, with around 9.5 million and 5 million customers, respectively. To date, the top four market players have all offered MNO-led mobile money services.

Mobile Connect

In February 2014, Groupe Speciale Mobile Association (GSMA) announced the launch of a collaborative initiative, supported by leading mobile operators (including Axiata Group Berhad, China Mobile, China Telecom, Etisalat, KDDI, Ooredoo, Orange, Tata Teleservices, Telefónica, Telenor, Telstra, and VimpelCom), to develop an innovative, interoperable ID that would allow consumers to securely access a wide array of digital services using their mobile-phone account for authentication (GSMA 2014). The first interoperable beta trial of Mobile Connect was launched by Dialog and Mobitel companies in Sri Lanka jointly in July 2014. WSO2, another Sri Lanka-based company, provided the Identity Server on OpenID Connect protocol, thereby providing broad interoperability across the mobile operators and online service providers.

The result is the world's first multioperator Mobile Connect solution that provides an out-of-band medium for authenticating a user to any service provider without requiring a password. The identity given under Mobile Connect is a form of a "validated" identity. The user's credentials are always stored with the "home" entity (the "identity provider"). When the user logs into a service, instead of providing credentials to the service provider, the service provider trusts the identity provider to validate the credentials. So the user never provides credentials directly to anybody but the identity provider.

The User Case

The mobile operators usually have a subscriber base with their identities already verified. Since Sri Lanka has unique national ID cards and SIM card registration is mandatory, mobile operators already have the customers' credentials. At the beta stage, popular online e-commerce sites and an insurance provider were linked to the system. The consumers were able to directly access the online sites via mobile phones.

The system provides two types of authentications: implicit, where the customer is logged into the website seamlessly; or explicit, where the user is prompted to click OK. If the user is not on a mobile network (Wi-Fi access or on a computer or other devices), the user will then be prompted to enter the mobile number. However, this information is viewed only by the identity provider and not by the online service provider. In both instances the service provider only receives a verification of the credentials.

Figure 6.7 Mobile Connect Beta Trial Indicators

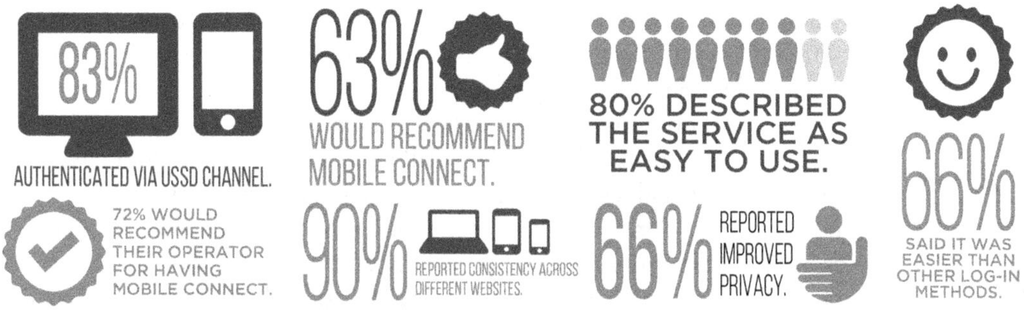

Source: GSMA 2015. © Groupe Speciale Mobile Association (GSMA). Reproduced, with permission, by GSMA; further permission required for reuse.
Note: USSD = unstructured supplementary service data.

The customers enjoy a hassle-free, consistent experience across different online sites and never need to remember multiple usernames or passwords. Moreover, the identity information is protected, and customers can surf the web and make online purchases and payments without having to worry how their personal information will be used by the service providers. Although the beta trial was not considered statistically significant and the results were taken only as indicative, the beta users acknowledged satisfaction with the service provided by Mobile Connect (figure 6.7).

Future Pathways

In Sri Lanka Mobile Connect is now available as a service for the entire customer base of both Dialog and Mobitel, totaling more than 15 million subscribers, and the companies are actively mobilizing service partners. Joint marketing and public awareness campaigns are ongoing. Dialog Axiata and Sri Lanka Telecom Mobitel won a special award at the European Identity Conference 2015, for the Mobile Connect implementation with the WSO2 Identity Server (*BusinessWire* 2015.)

By 2015, the GSMA had already announced that 17 MNOs had launched the Mobile Connect service in 13 countries, with plans for additional launches and beta trials to follow. The GSMA's Mobile Connect service enables customers to create and manage a universal identity that will securely authenticate them and allow them to safely access mobile and digital services such as e-commerce, banking, health, and digital entertainment as well as e-government portals via their mobile phones.

Because digital identity reduces the uncertainties inherent in transactions effected from a distance, using digital identity has huge potential for developing countries. Given the proliferation of mobile phones in developing countries, mobile identity can be used effectively in a variety of ways. There are different levels of security, from low-level website access to highly secure bank-grade authentication. This can be effectively used by government welfare schemes, for remittance transactions, financial transactions, and other online e-commerce sites. Even for poor and older people who may find navigating the complex digital world challenging, having a digital identity through an identity provider would work more effectively than having to remember usernames, personal identification numbers (PINs), and passwords.

One very important issue to be addressed before scaling up this service is an evaluation of the e-identity laws and regulations in a country. Even Sri Lanka does not have such laws and regulations. Such regulations and guidelines would increase the effectiveness and transparency of public and private sectors' electronic services by providing legal clarity in terms of security and data protection requirements, national supervisory mechanisms, and reinforced accountability and interoperability frameworks. Hence, developing countries should look at developing digital identity services in a holistic manner.

Notes

1. A single country may have multiple cases (up to 15, for India) involving different actors at the state, local, and national levels. The total population being targeted is estimated.
2. iPhone 5S and Samsung Galaxy S5 and subsequent-generation phones come with fingerprint scanners.
3. BSP Circular No. 608 Series of 2008.
4. Sari-Sari ("variety") are little convenience stores.
5. Under the "business correspondent" model, nongovernmental organizations/microfinance institutions set up under Societies/Trust Acts, Societies registered under Mutually Aided Cooperative Societies Acts or the Cooperative Societies Acts of States, section 25 companies, registered nonbanking finance companies not accepting public deposits, and Post Offices may act as business correspondents (RBI 2006).
6. The DBT of LPG scheme, PAHAL (Pratyaksh Hanstantrit Labh).

Bibliography

BFA (Bankable Frontier Associates). 2010. "Demand Study of Domestic Payments in the Philippines." Report for the Bill and Melinda Gates Foundation, BFA, Somerville, MA.

BSP (Bangko Sentral ng Pilipinas). 2014. "Status Report on the Philippine Financial System – 1st Semester 2014." BSP, Manila.

BusinessWire. 2015. "WSO2 Customer Dialog Axiata and Mobitel Receive Special Award at the European Identity and Cloud Awards 2015." May 21. http://www.businesswire.com/news/home/20150521005375/en/WSO2-Customer-Dialog-Axiata-Mobitel-Receive-Special.

Chakrabarty, K. C. 2014. "Currency Management in India: Issues and Challenges." Keynote address at the Banknote Conference 2014, Washington, DC, April 8.

Economic Times. 2013. "UID Project: UID Number Enrolment." *IndiaTimes: The Economic Times*, August 23.

———. 2014. "I-cards Cost under Aadhaar World's Cheapest, Says DeitY Secretary." *IndiaTimes: The Economic Times*, February 12.

Finextra. 2014. "Mobile Operators Band Together to Create Interoperable Digital ID." March 25. https://www.finextra.com/news/announcement.aspx?pressreleaseid=54547&topic=security.

Gelb, A., and J. Clark. 2013. "Identification for Development: The Biometrics Revolution." Center for Global Development (CGD) Working Paper 315, CGD, Washington, DC.

GSMA (Groupe Speciale Mobile Association). 2013. "The Mandatory Registration of Prepaid SIM Card Users." White paper, GSMA, London.

———. 2014. "Leading Mobile Operators Unveil GSMA Mobile Connect Initiative to Provide Consistent and Interoperable Approach to Managing Digital Identity." GSMA News, February 24.

———. 2015. "Mobile Connect Beta Trial." Research paper, GSMA, London.

Harpur, P. 2016. "Sri Lanka—Telecoms, Mobile and Broadband—Statistics and Analyses." Report on BuddeComm telecommunications research website (updated September 21): https://www.budde.com.au/Research/Sri-Lanka-Telecoms-Mobile-and-Broadband-Statistics-and-Analyses.

Hickok, E. 2013. "Unique Identification Scheme (UID) and National Population Register (NPR), and Governance." Background note, Centre for Internet and Society (CIS), March 14.

IMF (International Monetary Fund). 2013. *Asia and Pacific: Shifting Risks, New Foundations for Growth*. Regional Economic Outlook, World Economic and Financial Surveys Series. Washington, DC: IMF.

ITU (International Telecommunication Union). 2012. *Measuring the Information Society 2012*. Geneva: ITU.

Mehra, P. 2015. "Budget for Huge Increase in DBT." *The Hindu*, March 3.

MGNREGA (Mahatma Gandhi National Rural Employment Guarantee Act). 2014. "Action Plan for Linking Aadhaar with MGNREGA: Strategy for Aadhaar Seeding in the 300 DBT Districts as a First Step for Introducing Aadhaar Enabled Payment System (AEPS) in MGNREGA." Action plan document, MGNREGA, Ministry of Rural Development, Government of India, New Delhi.

MPNG (Ministry of Petroleum and Natural Gas, Government of India). 2014. "Launch of PAHAL (DBTL) Scheme on 1st January 2015 in Entire Country." Press release, December 31.

NIPFP (National Institute of Public Finance and Policy). 2012. "A Cost-Benefit Analysis of Aadhaar." Study report, NIPFP, New Delhi.

Rao, K. V. 2013. "RBI Notifies eKYC as Valid." *livemint* (e-paper), September 4.

RBI (Reserve Bank of India). 2006. "Financial Inclusion by Extension of Banking Services—Use of Business Facilitators and Correspondents." Circular RBI/2005-06/288DBOD.No.BL.BC. 58/22.01.001/2005-2006, RBI, Mumbai.

Srikanth, R. 2013. "A Study on—Financial Inclusion—Role of Indian Banks in Reaching Out to the Unbanked and Backward Areas." *International Journal of Applied Research and Studies* 2 (9).

Tewari, R. 2014. "UPA's Aadhaar, DBT Schemes Get Boost from PM Modi." *Indian Express*, July 20.

UIDAI (Unique Identification Authority of India). 2012a. "Aadhaar Enabled Service Delivery." White paper, UIDAI Planning Commission, Government of India, New Delhi.

———. 2012b. "Role of Biometric Technology in Aadhaar Enrollment." Report, UIDAI, New Delhi.

———. 2016. "AADHAAR E-KYC API Specification - Version 2.0." UIDAI, New Delhi.

UNHCR (United Nations High Commissioner for Refugees). 2017. "Ending Statelessness." UNHCR website: http://www.unhcr.org/stateless-people.html.

UNICEF (United Nations Children's Fund). 2013. "Every Child's Birth Right: Inequities and Trends in Birth Registration." Report, UNICEF, New York.

PART III

South Asia Digital Landscape, Future Options, and Conclusions

This study has examined how macro-, meso-, micro-, and customer-level actors can lead, engage, and influence the policies, regulations, processes, and ecosystems to achieve greater financial inclusion through critical game-changing measures that foster innovative digital solutions. The experiences of the countries studied, both within and outside the South Asia region, have revealed many useful lessons that can be used as guiding principles to enhance financial inclusion in South Asia and in the developing world as a whole.

Part III aims to distill the guiding principles and measures that would help avoid the deficiencies and pitfalls that have constrained the pace of scaling-up in several countries, and to suggest (in chapter 8) how they could be applied to help change the digital landscape in South Asia (summarized in chapter 7) as well as in developing countries more generally (chapter 9). Even though the game-changing elements discussed in this study, and digital finance per se, cannot be considered a panacea for persistent financial exclusion problems, the lessons discussed here provide valuable options for policy makers and market players in addressing shortfalls in the use and reach of financial services.

CHAPTER 7

Digital Landscape in South Asia

Introduction

South Asia consists of eight countries at different stages of economic development, with widely varying economic characteristics and attributes. With nearly 40 percent of the developing world's poor living in the region (World Bank 2015) and 54 percent of the region's adult population (585 million) financially excluded,[1] addressing the financial inclusion gap is critically important.

Comparison of Findex data across the regions reveals that the South Asia region lags behind the East Asia and Pacific, Europe and Central Asia, and Latin America and the Caribbean regions in the percentage of adults with accounts, females with accounts, and the poorest 40 percent with accounts in formal financial institutions (table 7.1). On the other hand, the percentage of young adults (ages 15–24 years) with accounts in South Asia compares well with the other regions.

Even though the mobile account ownership indicator for South Asia is the second highest behind Sub-Saharan Africa, at 2.6 percent, considerable efforts are needed to push the mobile account ownership numbers closer to the 11.5 percent achieved in Sub-Saharan Africa. South Asia shows limited ownership and use of debit cards and credit cards (8.5 percent and 2.6 percent, respectively). Although automated teller machines (ATMs) seem to be the preferred method of cash withdrawal (31.1 percent), the region scores relatively low on receipts of wages and government grants to accounts (3.5 percent and 3.1 percent, respectively) and utility bill payments using an account (2.7 percent). These gaps represent opportunities to use digital means to reach the poor and unbanked in a more meaningful manner.

Before examining the applicability of the lessons and guiding principles from the case study countries, it is important to diagnose the digital landscape and readiness of each country. Detailed country reports based on publicly available information and data gathered from different sources are summarized in an "at-a-glance" matrix in chapter annex 7A, table 7A.1. This chapter also reviews the status of South Asian countries in terms of the key themes at different levels of analysis, as a basis for applying the lessons of experience and guiding principles in chapter 8.

Table 7.1 Financial Inclusion Data by Region, 2014
Percentage

Data type	*East Asia and Pacific*	*Europe and Central Asia*	*High-income OECD*	*Latin America and the Caribbean*	*Middle East*	*South Asia*	*Sub-Saharan Africa*
Account (age 15+) (w2)	69.0	51.4	94.0	51.4	14.2	46.4	34.2
Account, female (age 15+) (w2)	67.0	47.4	93.8	48.6	9.2	37.4	29.9
Account, income, poorest 40% (ages 15+) (w2)	60.9	44.2	90.6	41.2	7.3	38.1	24.6
Account, young adults (ages 15–24) (w2)	60.7	35.6	84.1	37.4	7.6	36.7	25.9
Account, rural (age 15+) (w2)	64.5	45.7	93.8	46.0	10.7	43.5	29.2
Account at a financial institution (age 15+) (w2)	68.8	51.4	94.0	51.1	14.0	45.5	28.9
Account at a financial institution (age 15+) (w1)	55.1	43.3	90.0	39.3	10.9	32.3	23.9
Mobile account (age 15+) (w2)	0.4	0.3	—	1.7	0.7	2.6	11.5
Debit card (age 15+) (w2)	42.9	36.9	79.7	40.4	8.5	18.0	17.9
Debit card (age 15+) (w1)	34.7	36.4	61.9	28.9	5.5	7.2	15.0
Main mode of withdrawal: ATM (with an account, age 15+) (w2)	53.3	66.7	—	71.1	44.9	31.1	53.8
Main mode of withdrawal: ATM (with an account, age 15+) (w1)	37.0	72.5	68.5	57.0	42.4	16.9	51.7
Used an account to receive wages (age 15+) (w2)	15.1	22.5	44.3	18.0	3.3	3.5	7.3
Used an account to receive government transfers (age 15+) (w2)	8.1	7.3	17.2	9.0	0.9	3.1	3.8
Used an account at a financial institution to pay utility bills (age 15+) (w2)	11.8	12.5	61.1	6.3	0.2	2.7	2.8
Debit card used in the past year (age 15+) (w2)	14.8	22.9	65.3	27.7	3.3	8.5	8.7
Credit card used in the past year (age 15+) (w2)	10.8	14.9	46.7	18.0	1.5	2.6	1.9

Source: Global Findex 2014 data, http://datatopics.worldbank.org/financialinclusion/.
Note: — = not available; ATM = automated teller machine; OECD = Organisation for Economic Co-operation and Development. "w1" denotes 2011 Global Findex data (wave 1), and "w2" denotes 2014 Global Findex data (wave 2).

Macro-Level Strategies

At the macro level (of policy makers, regulators, and donors), a few of the South Asian countries have started aligning policies and strategies on financial inclusion as a national agenda. Lack of a clear national strategy—as well as changes in strategy when governments change—has resulted in inconsistent policies and discontinuity of results-oriented programs. The region's countries vary widely in their approaches, although several have made progress:

- *India* now has a clear direction and vision, and the country has embarked on a financial inclusion strategy focused heavily on digital financial inclusion modalities.
- *Pakistan* has not yet formulated a comprehensive digital agenda for financial inclusion, although its high-value payment system is digitized, and it reportedly launched a national financial inclusion strategy in 2015.
- *Sri Lanka* does not have a national-level, comprehensive financial inclusion strategy, but the country has a clear information and communication technology (ICT) and e-government policy, and the Central Bank has pursued use of digital finance for financial inclusion under its payment systems policy agenda.
- *Bangladesh* is also driving financial inclusion as part of its strategic plan, and it has a national ICT policy that promotes digitization.

Meanwhile, a comprehensive national policy or strategy on financial inclusion is absent in Afghanistan, Bhutan, Maldives, and Nepal, although efforts by their respective central banks to address financial inclusion issues can be observed in each country. More recently, the Maldives Monetary Authority (MMA), together with the Ministry of Finance and Treasury, has decided to use mobile network operator (MNO)-led mobile money as a short-term solution to address access to finance and financial inclusion issues in the outer islands and atolls. In Nepal, a large number of poverty alleviation–related programs have been initiated with the help of donor agencies, but there is no comprehensive program for the use of digital finance as a solution to address issues relating to financial inclusion. Post-earthquake Nepal is experiencing many new issues and may need to draw up a long-term strategy for this purpose. Afghanistan has a comprehensive wish list, but it still has to prioritize the main programs that would address financial inclusion issues. Bhutan is moving slowly with a declaration to support financial inclusion, but it has yet to implement a sustainable strategy.

Traditionally, South Asian central banks have been the driving force in financial sector development. However, lack of appropriate legal and regulatory structures has been a common problem of the e-money landscape in South Asia. Nonbanks have taken the lead in providing e-money services at an affordable cost to the poorer groups. The phenomenal growth in the use of mobile phones in South Asia means that MNOs have the capacity to surpass the traditional banking model in terms of facilitating access by the poor and unbanked to basic payment and financial services.

However, because MNOs are outside the ambit of the financial regulatory authorities, special dispensations are often necessary to permit these nonbanks and MNOs to enter the payment space to provide mobile-phone–based e-money solutions to the poor and the unbanked. India, Pakistan, and Sri Lanka have enacted Payment and Settlement Laws. The other South Asian countries have no payment laws but have used the provisions in the central bank or the monetary authority Acts to issue relevant regulations for this purpose. Bangladesh has drafted a Payment Law, and Maldives has submitted a bill to the parliament. All countries have Anti-Money Laundering and Combatting Funding of Terrorism (AML/CFT) laws or regulations covering financial transactions, many of which are being reviewed.

Among the South Asian countries, Sri Lanka has the most comprehensive set of enabling laws and regulations for digital finance and facilitating nonbanks to enter into the payment space. It continues to adopt a flexible approach toward promoting nonbank entry, but within the applicable legal framework.

Meso-Level Approaches and Issues

A supportive infrastructural environment for e-money is important for digital finance to be successful as a means of enhancing financial inclusion. However, many South Asian countries continue with ad hoc, proprietary, stand-alone systems that do not necessarily provide a low-cost, affordable solution for the customers.

India has begun the process of making formal financial services available at affordable costs—for example, by mandating its banks to use a common application programming interface (API) for mobile banking and ensuring that bank-led e-money projects are interoperable at all possible levels. The country's biometric identification (ID) initiative, Aadhaar, is an exceptional measure India has introduced to ensure that a unique ID is available to help mobile-phone users to open accounts and social grant recipients to access funds anywhere in the country. Financial infrastructure provision is well coordinated by the Reserve Bank of India, with support from the relevant ministries.

For its part, Sri Lanka has put key financial infrastructure in place, and systems are running well to ensure that services are available. However, coordination at the policy and regulatory levels needs improvement, in particular to address the passive role played by the national payments committee (NPC). LankaClear Pvt. Ltd., the retail payment infrastructure provider, is ready to move faster toward a fully interoperable Common Switch, but this requires tight coordination among stakeholders.

Bangladesh is still hesitant to allow MNOs to enter the payments arena, despite the problems encountered in the bank-led agent network. Maldives is a difficult geographical terrain for rolling out the necessary financial infrastructure, and the outer islands and atolls have no efficient and conveniently available access to payment and financial services. Therefore, a mobile money initiative is now under way in Maldives that will be low-cost relative to services from banks and other financial institutions.

Achieving full-scale financial inclusion is difficult in both Pakistan and Afghanistan because of ongoing security situations. Nevertheless, authorities are taking steps to deal with operational issues while promoting user-friendly digital products to enhance financial inclusion. Nepal has a private switch and a national common switch, but a comprehensive national strategy is needed to drive efficiency. Interoperable National Payment Switches (Common Switch) have been established in Bangladesh and Sri Lanka, although not all banks have joined these national switches.

One unique feature about the South Asia region is that all eight countries are members of the South Asian Association for Regional Cooperation (SAARC) Payment Council (SPC), established in 2008 to promote sound payment and settlement systems in the member countries. Digital finance as a means to address financial inclusion issues can be taken up by this regional body.

Agent network management is a concern for mobile money in most South Asian countries. Most MNO-led or bank-led but agent-managed initiatives are experiencing issues with respect to over-the-counter (OTC) transactions—that is, transactions that the agent conducts on behalf of a sender/recipient or both from either the sender's or the agent's mobile money account. These OTC transactions are often preferred by illiterate customers because they are apprehensive about effecting their own transactions using their mobile or prepaid cards. However, these agent-assisted transactions pose a number of issues including limiting the evolution of a digital cash-lite ecosystem; decreased provider profitability because of the high cost of monitoring; and unregistered transactions, which run the risk of money laundering and terrorism financing.

In Sri Lanka OTC is not encouraged. In Bangladesh, the bKash system suffers from liquidity rebalancing and consumer protection issues, as customers often share their pin numbers with the agents. These issues require improved governance at the agent level, and financial education at the customer level.

India also suffered a long time with OTC-type transactions, but more recently has attempted to address some of the recurring issues. An MNO agent system is cheaper than the bank-led models' agent system, and it is critical for MNO operations. Often, these agents are nondedicated to the mobile money operation, and they engage in many other ancillary services.

Micro-Level Models

South Asian countries are heavily bankcentric, and financial system development is usually driven through regulations. Although countries have recognized the financial inclusion gaps and tried to find solutions, most of them still favor bank-led models over MNO-led mobile money solutions.

India passed legislation permitting "payment banks" and "small finance banks," and deposit accounts in those entities are eligible for risk-proportionate know-your-customer (KYC) procedures. The Reserve Bank of India (RBI) also amended the regulation enabling nonbank entities to be banking correspondents. This will enable nonbank entities, including mobile operators, and the national postal service (with a larger network) to offer financial services by obtaining a banking license. However, these payment banks and small banks will be subjected to prudential regulations. Hence, the minimum capital requirement and other prudential regulations will need to be addressed to facilitate scaling up to reach the masses. And in Bangladesh, bKash—a fully owned subsidiary of BRAC Bank with more than 80,000 agents—is the second largest mobile money company in the world in terms of the number of individual accounts.

Only Afghanistan and Sri Lanka have allowed MNOs to enter the payment space and offer MNO-led mobile money services. All others have stayed with the bank-led model or some variation of it, partly because of legal and regulatory issues. Although MNOs have the natural advantage of reaching rural people and the poor, as they already have their last mile touchpoints in place, South Asian policy makers and regulators generally remain hesitant to promote MNO-led mobile financial services. Nevertheless, Maldives is actively launching an MNO-led mobile money solution.

Sri Lanka has the world's first single-wallet, end-to-end interoperability among three MNOs. In terms of ATM interoperability, Bangladesh, India, Pakistan, and Sri Lanka have full or partial interoperability. Most significant is India's Aadhaar-enabled accounts that allow for interoperable transactions through any banking business correspondent of any bank using Aadhaar biometric verification and authentication. If successful, this would truly be a game changer for financial inclusion. Some forms of interoperability exist in Nepal (eSewa payment gateway) and Pakistan (between MNOs and banks).

South Asian countries have not used digital means to provide government salaries or social grants in an effective manner. Afghanistan used mobile money to send civil servant salaries, and India recently launched Aadhaar-enabled Jan Dhan Yojana bank accounts to channel direct benefit transfers and all government payments. Digitizing government-to-person payments has already started in India by depositing government pension and scholarship payments directly into bank accounts in some of the districts. Pakistan has also used National Database and Registration Authority (NADRA) prepaid biometric cards to disburse flood relief to victims, but it has yet to configure all grant disbursements to go through such a mechanism. All others use a combination of cash and savings accounts in microfinance institutions or commercial banks to disburse government benefits. There is a huge opportunity to streamline grant payments, government salaries, and pension payments.

Customer-Level ID Systems

The lack of a unique identity has been a continuing problem in the South Asia region, and these countries will need to introduce digital ID systems to enable the lower-income population to access formal financial institutions. India now has provided a long-term solution: the Unique Identification Authority of India (UIDAI)'s Aadhaar program is continuing the process of assigning ID numbers to all citizens, having already issued 855 million IDs (68 percent of the population).

Afghanistan, Bhutan, Maldives, and Sri Lanka have unique IDs, and Sri Lanka is also planning to go for a biometric ID system that would solve many issues related to fake IDs submitted at the time of opening a bank or transactional accounts. Afghanistan came out with a digital ID (electronic-Tazkera) in 2015. Bangladesh, Bhutan, and Nepal have multiple IDs, as does Pakistan—although it now offers a Computerized National Identity Card (CNIC) by NADRA that verifies users' biometric data. NADRA also does biometric verification when registering a new subscriber identification module (SIM) card.

Although all South Asian countries have data on grant recipients, India and Pakistan have developed biometric-enabled databases, which is a significant step forward. South Asia could learn from South Africa's successful experience with a biometric ID system (as discussed in chapter 6), which has managed to avoid some of the KYC and customer due diligence and AML/CFT issues.

Annex 7A Digital Financial Landscape in South Asia, by Country: At a Glance

Table 7A.1 Digital Financial Landscape in South Asia: At a Glance

Digital finance enabler	*Afghanistan*	*Bangladesh*	*Bhutan*	*India*	*Maldives*	*Nepal*	*Pakistan*	*Sri Lanka*
A. MACRO-LEVEL POLICIES, STRATEGIES, LAWS, AND REGULATIONS								
1. Declared national policy or strategy								
National financial inclusion policy or strategy available	No	No separate policy; Central Bank drives financial inclusion strategy (Bangladesh Bank's strategic plan 2010–14 strategy #4) National financial inclusion strategy being drafted	No: Maya Declaration signed; financial inclusion policy drafted	Yes: Pradhan Mantri Jan Dhan Yojana, 2014	No	Demand survey done under MAP program UNCDF; on track to prepare road map to financial inclusion	Yes: financial inclusion strategy, 2015	No: CBSL drives the financial inclusion strategy
National digital financial agenda	No: Commitment to Better than Cash Alliance; government-formed Digital Finance Committee to oversee the salary payments of civil servants	Yes: Digital Bangladesh Vision 2021	No	Yes	No	No	Yes: as part of financial inclusion strategy	No explicit agenda
National ICT policy or e-government policy	No	Yes: Digital Bangladesh Vision 2021	No	Yes	No	No	Draft national ICT policy of 2012	Yes: e-government

table continues next page

Table 7A.1 Digital Financial Landscape in South Asia: At a Glance *(continued)*

Digital finance enabler	*Afghanistan*	*Bangladesh*	*Bhutan*	*India*	*Maldives*	*Nepal*	*Pakistan*	*Sri Lanka*
National payment strategy	No	Being done	Being done	Yes	No	No	Not clear whether there is a long-term strategy	No: need to assess the long-term strategy
2. Legal and regulatory (enabling laws and regulations)								
Payment and settlement law	No	No (draft act)	No	Yes	Bill stage	No	Yes, but restrictive	Yes
Payment regulations under other laws (central bank law, monetary authority law, or banking act)	Yes	Yes	Yes	Yes	Yes	Yes: NRB IT Policy, #2068 under NRB Act	Yes	Yes
AML/CFT law or financial transaction law or regulations	Yes	Yes	Yes	Yes	Yes	Yes	Yes	Yes
Financial consumer protection law or regulations	No	No	No: through other acts and regulations	Yes: also banking ombudsman	No	No	No	No: only through financial ombudsman and CBSL mechanisms
Electronic funds transfer law or regulations	Yes: e-money issuer regulations 2011	Yes	No	Yes: regulations under RBI Act	No	No	Yes	Yes

table continues next page

Table 7A.1 Digital Financial Landscape in South Asia: At a Glance *(continued)*

Digital finance enabler	*Afghanistan*	*Bangladesh*	*Bhutan*	*India*	*Maldives*	*Nepal*	*Pakistan*	*Sri Lanka*
A. MACRO-LEVEL POLICIES, STRATEGIES, LAWS, AND REGULATIONS *(continued)*								
2. Legal and regulatory (enabling laws and regulations) *(continued)*								
Electronic or computer transactions law or digital finance law	No	No	No	India IT Act 2000, IT (Amendment) Act 2008 deals with electronic offenses	No	No	No	Yes
Payment devices fraud law	No	No	No	Same as above	No	No	No	Yes
Computer crimes law	No	No	No	Same as above	No	No	No	Yes
B. MESO-LEVEL SYSTEMS AND NETWORK CHARACTERISTICS								
National payment system	Work ongoing with APS	Yes	Yes	Yes	MMA operates automated clearinghouse; BML operates the card and proprietary ATM switch	ATM switch, SCT switch, and Payway gateway	Yes	Yes
National financial infrastructure provider (retail payments)	Work ongoing with APS	Yes: Bangladesh common payment switch	Yes: Royal Monetary Authority of Bhutan EFTCS	National Payments Corporation of India			National Institutional Facilitation Technologies (Pvt.) Ltd. 1LINK and Mnet interoperable ATM switches	Yes: LankaClear (Pvt.) Ltd.
ICT agency	No	No	Yes	Yes	No	No	No	Information and Communication Technology Agency
Industry involvement	No	Unclear: BASIS might be involved in financial sector digitization	Unclear: BICTTA might be involved in financial sector digitization	NASSCOM	No	No	Possibly: National ICT R&D Fund	SLASSCOM

table continues next page

Table 7A.1 Digital Financial Landscape in South Asia: At a Glance *(continued)*

Digital finance enabler	*Afghanistan*	*Bangladesh*	*Bhutan*	*India*	*Maldives*	*Nepal*	*Pakistan*	*Sri Lanka*
Credit information bureau and secured transaction registry	Public credit registry Unified collateral registry However, not fully operationalized?	Credit information bureau Online platform No collateral registry	Credit registry No collateral registry	Credit registry and collateral registry	Public credit registry No collateral registry	Private credit information bureau Secured transaction law available; registry not implemented	Public credit registry No collateral registry	Yes: credit information bureau and electronic searchable collateral registry (need amendment to law)
Regional policy advocates or networks Regional payment council (SAARC)	Member	Member	Member	Member	Member	Member	Member	Member
Common national payment switch available?	No	Yes, being enhanced	Yes	Yes	No	Yes: three switches	No	Yes
ATM network available?	Yes: mainly in Kabul city	Yes	Yes	Yes	Yes	Yes	Yes	Yes
ATM network interoperable?	No, but encouraged	Partially	No	Yes	No	No	Yes	Yes
EFTPOS network available?	Some	Yes	Yes	Yes	Yes: BML	Yes	Yes	Yes
MNOs providing services on behalf of micro-level players	Yes	Yes	Yes	Yes	Yes	Yes	Yes	Yes

table continues next page

Table 7A.1 Digital Financial Landscape in South Asia: At a Glance *(continued)*

Digital finance enabler	*Afghanistan*	*Bangladesh*	*Bhutan*	*India*	*Maldives*	*Nepal*	*Pakistan*	*Sri Lanka*
B. MESO-LEVEL SYSTEMS AND NETWORK CHARACTERISTICS *(continued)*								
Nonbank-led agent networks available?	Yes: no agent banking regulations	Yes: on behalf of bank-operated mobile money solutions	No: just a network of agents who are Bank of Bhutan agents	Yes	No	Yes, but guidelines not clear	Yes	Yes
Custodian bank or trusteeships available?	Yes	Yes	No	No	Yes	No	No	Yes
C. MICRO-LEVEL MODELS, SYSTEMS, AND SERVICES								
Mobile money: bank-led model available?	Yes, some banks do	Yes	Yes	Yes	Yes	Yes	Yes	Yes
Mobile money: nonbank-led model available?	Yes	No: bank subsidiary-led comes under bank-led model	No	No: now all have to get payment bank status	Yes: just launched	No	No	Yes
MNO-led model available?	Yes	No	No	No	Yes	No	No	Yes
Number of nonbank-led mobile money deployments	4	9	0	15	2	3	7	2 (3 MNOs interoperate one wallet)
Interoperability among nonbank-led mobile money initiatives?	No, but encouraged	No	No	No	No: planned	Yes: in eSewa, a payment gateway	Yes: mobile to bank	Yes

table continues next page

Table 7A.1 Digital Financial Landscape in South Asia: At a Glance *(continued)*

Digital finance enabler	*Afghanistan*	*Bangladesh*	*Bhutan*	*India*	*Maldives*	*Nepal*	*Pakistan*	*Sri Lanka*
Services provided by nonbank-led mobile money initiatives, including OTC	Cash-in/cash-out, bill pay, merchant pay, remittance, salary payments, MFI loan disbursement or repayment Airtime top-up, link to other banking products	Merchant or bill payment Other bulk payment P2P transfer (domestic) International remittances (inward only) Airtime top-up, G2P, B2P Link to other banking products	None	Airtime top-up Link to other banking products Bill payment Other bulk payment P2P transfer (domestic) (depending on service provider)	Bill/merchant payment P2P transfer (domestic) Airtime top-up	Wallet-to-wallet transfers Wallet top-up from a bank account Wallet-to-merchant and online payments Prepaid cards for wallet top-up	Merchant payment Airtime top-up Loan disbursement or repayment International remittances P2P transfer (domestic) Other bulk payment Bill payment Mobile microinsurance	Merchant payment P2P transfer (domestic) Airtime top-up Bill payment Depending on provider, NFC to bus services, international remittance, government pension payments
Mobile money applications?	No	No	No	No	No	No	No	NFC-enabled remittance app, Mobile Connect ID
Social grant disbursement agency	Banks and financial institutions Government pensions and national solidarity program	State-owned commercial banks, MFIs	Commercial banks	Yes: Aadhaar-enabled Jan Dhan Yojana accounts and state bank	BML (state-owned commercial bank)	State-owned commercial banks	Banks, MFIs	State banks. Samurdhi, and Divineguma banks
Method of social grant disbursements	Bank accounts, MFIs	Bank account, BEFTN, MFS, cards	Accounts, cash	Cash or bank account or cards?	BML accounts or cash	Account or cash	Cash or bank mobile-phone banking or prepaid cards	Bank accounts

table continues next page

Table 7A.1 Digital Financial Landscape in South Asia: At a Glance *(continued)*

Digital finance enabler	*Afghanistan*	*Bangladesh*	*Bhutan*	*India*	*Maldives*	*Nepal*	*Pakistan*	*Sri Lanka*
***C. MICRO-LEVEL MODELS, SYSTEMS, AND SERVICES** (continued)*								
Digitized social grant disbursements	No	No	No	Yes: started	No: planned	No	Yes: NADRA rolled out prepaid flood relief cards, but not all disbursed through this manner	No: project underway
If digitized, is it biometric enabled?	No	No	No	Yes	No	No	Yes, for the flood-relief cards	No
D. CUSTOMER-LEVEL ID POLICIES OR SYSTEMS								
Unique ID?	Tazkera (national ID)	Several forms of ID	Yes	Yes: Aadhaar and National Population Register	Yes: national ID	Voter registration and many other IDs	Computerized National Identity Card	Yes: national ID
Digital ID?	No	No	No	Yes	No	No	Yes	Yes: Mobile Connect in smartphones
Biometric ID?	Yes: electronic: Tazkera began August 2015	No	No	Yes	No	No	Yes	No
Social grant recipient database?	Yes?	Yes?	Yes	Yes	Yes	No	Yes	Yes

Note: AML/CFT = Anti-Money Laundering and Combatting Funding of Terrorism; APS = Afghanistan Payments Systems; ATM = automated teller machine; BASIS = Bangladesh Association of Software and Information Services; BEFTN = Bangladesh Electronic Funds Transfer Network; BICTTA = Bhutan Information and Communication Technology Training Association; BML = Bank of Maldives; B2P = business-to-persons; CBSL = Central Bank of Sri Lanka; EFTCS: Electronic Funds Transfer and Clearing System; EFTPOS = electronic funds transfer at point of sale; G2P = government-to-persons; ICT = information and communication technology; ID = identification; IT = information technology; MAP = Making Access Possible; MFI = microfinance institution; MFS = Mobile Financial Services; MMA = Maldives Monetary Authority; MNO = mobile network operator; NADRA = National Database and Registration Authority; NASSCOM = National Association of Software and Services Companies; NFC = near-field communication; NRB = Nepal Rastra Bank; OTC = over-the-counter; P2P = person-to-person; SAARC = South Asian Association for Regional Cooperation; SCT = SmartChoice Technologies; SLASSCOM = Sri Lanka Association of Software and Service Companies; UNCDF = United Nations Capital Development Fund.

Note

1. Data from the Global Findex Survey 2014, http://datatopics.worldbank.org/financialinclusion/.

Bibliography

World Bank. 2015. *Global Monitoring Report 2014/2015: Ending Poverty and Sharing Prosperity*. Washington, DC: World Bank.

CHAPTER 8

Opportunities, Challenges, and Future Options in South Asia

Broader adoption of digital payments—with regard to both remittances and other payments—can significantly advance the global financial inclusion agenda and support the priority areas of the Global Partnership for Financial Inclusion (GPFI). Not only are digital payments more efficient than cash payments, but their broader adoption also can reduce rates of corruption and violent crime, reduce the cost of government wage and social transfer payments, offer new pathways into the financial system for the disadvantaged, and, importantly, contribute to the ongoing objective of women's economic empowerment.

—Leora Klapper and Dorothe Singer,
"The Opportunities of Digitizing Payments"

Introduction

While the previous chapter's summary of the digital landscape in South Asia cannot replace an in-depth diagnostic assessment of how best to digitize the delivery of financial services in each country, it provides a basis for applying the lessons and principles that emerged from the analysis provided in Part II. The case study countries provided insight into how game-changing critical enablers can enhance the effectiveness of, or help overcome barriers to, digital financial interventions at the macro, meso, micro, and customer levels. South Asia region countries can benefit from strong actions by the key stakeholders to give priority to these critical enablers and provide the vision, direction, regulatory adaptations, and resources to ensure their effective implementation.

Macro Level

Leadership and National Policies

National-level policy commitment, progressive leadership, and consistency are important enablers. It is an imperative for governments, policy makers, and regulators to clearly announce the objectives of policies for financial inclusion and e-money, as well as the specific areas of focus for defined time periods. Implementation authorities may simply ignore vague or unclear public policy statements or be stymied by changes in policy when the government changes. Countries should have a national policy for financial inclusion and a digital financial agenda (including information and communication technology [ICT] policy and e-government) to ensure that the strategy is implemented broadly to reach the unbanked and the underbanked.

Most South Asian countries have yet to develop these strategies. Among the case study countries, India's Pradhan Mantri Jan Dhan Yojana financial inclusion strategy is a clear example of how commitment to a major national strategy can drive significant efforts to bring poor people into the formal financial system and use digital mechanisms to reach them in a cost-effective manner.

Enabling Regulatory Environment

South Asian central bankers and regulatory authorities have always been at the helm of financial sector development, with a fair share of success. However, some still hold on to old definitions of intermediation and access as bank-only concepts. With most capital markets in South Asia at nascent stages of development, authorities' understandable concern about financial sector stability pushes them to maintain the status quo. But rapid development of digital payments and other e-platforms have demonstrated that the approach to financial innovation and regulation needs to be recast if persistent poverty and financial exclusion are to be addressed more effectively.

Regulators in India and Sri Lanka (like those in Kenya) have started the process of establishing a more conducive and nonprohibitive policy, legal, and regulatory environment. The Central Bank of Sri Lanka proactively established the legal framework for nonbank institutions to enter the payment space long before the need arose, including legislation regarding payments, consumer protection, Anti-Money Laundering and Combatting Funding of Terrorism (AML/CFT), and electronic funds transfer.

By letting nonbanks into the payment space, the regulator establishes a level playing field in which competition fosters innovations that enhance the customer's value proposition. At the same time, the regulator can address various market developments and innovations in a more holistic manner. Sri Lanka has successfully maintained the delicate balance between establishing a level playing field among operators that encourages innovation while providing adequate risk-based oversight and supervision.

Nevertheless, South Asian countries should take precautions against the risk of cyberthreats when using digital finance for financial inclusion. Although the damage in terms of value may be small, cybercrimes can wipe out people's savings and exacerbate the already limited confidence of poorer and more vulnerable groups in the formal financial system. Hence the regulator's role is critical to ensure that nonbanks as well as banks abide by prudential regulations and security guidelines.

Meso Level

Financial Infrastructure Development

The widely dispersed poor population in many South Asian countries is a key challenge for establishing the infrastructure needed to expand financial inclusion. Connectivity, national switches, interoperability, and agent network management are critical components of digital finance models. Coordination among macro-level policy makers and regulators, meso-level institutions, and micro-level service providers is key.

National retail payment providers, ICT agencies, and credit information bureaus are some examples of meso institutions that can enable infrastructure development—yet these remain largely undeveloped in most South Asian countries. Since all eight South Asian countries are members of the South Asian Association for Regional Cooperation (SAARC) Payments Council, it should incorporate digital finance as a priority agenda for financial inclusion at the regional level.

Meso-level enablers are evident in several of the case study countries. In Thailand, the proactive Thai Bankers' Association helps drive the policy that provides for the spread of ubiquitous automated teller machines (ATMs) throughout the country. Sri Lanka's ICTA (ICT agency) drives the e-government program, and LankaClear (retail infrastructure provider) promotes the corporate theme of "A World beyond Cash" in moving toward a fully interoperable common payment switch. The forward-thinking National Payments Corporation of India (NPCI) facilitates strategic implementation of its dynamic inclusion policy.

Agent Network Management

The case that mobile network operators (MNOs) are best placed to offer mobile money is based on fast-growing mobile penetration among the poor and the unbanked population and their last mile coverage through ubiquitous agent networks that can act as mobile money touchpoints. Managing the agent network is a critical enabler for rapid mobile money deployment, through training in handling the customers, implementing know-your-customer (KYC) and customer due diligence (CDD) requirements, and addressing suspicious transactions.

Poor or unbanked people identify themselves more readily with local MNO agents than with formal bank outlets. South Asian countries have yet to develop such agent networks as effectively as in Kenya, where they facilitated the phenomenal growth of M-Pesa, which has now become the country's national payments system and platform for many innovations that promote financial inclusion even among the poor in remote areas.

Micro Level

MNO-Led Mobile Money Model

Despite its clear advantages in terms of proximity and services to the targeted poor and unbanked population in hard-to-reach rural areas and islands, the MNO-led model is still not popular in the South Asia region. Only Afghanistan and Sri Lanka operate full-fledged MNO-led mobile money solutions. Maldives has launched its own MNO-led mobile money solution. And Bangladesh has bKash, which is owned by BRAC Bank.

A recent Groupe Speciale Mobile Association (GSMA) article (di Castri 2013) concludes that regulators in a number of countries (including Colombia, Ghana, Guatemala, India, Nigeria, and South Africa) have been reluctant to grant MNOs mobile money licenses because they are (a) afraid that the MNOs could scale up very quickly and dominate the formal financial sector; (b) concerned that regulators would be unable to control such growth and use it to deepen access to finance; and (c) nervous about the ability of MNOs to safeguard customers' money as effectively as banks. But in many other markets, the regulator has been able to design prudential regulations and market conduct frameworks that help to make nonbank e-money providers sound and effective.

Interoperability

Interoperability is another important enabler for MNOs to enhance their outreach and efficiency. However, not many countries in South Asia (or elsewhere) have MNOs that interoperate. The best example is Sri Lanka, which launched the world's first end-to-end interoperable mobile money solution: three MNOs operate a single wallet sharing merchants and agents, with firewalls built in so they cannot access others' subscriber information. Indonesia represents another interoperability concept, with four MNOs interoperating but retaining four individual wallets. Through such mechanisms, customers stand to benefit from lower cost as well as a larger number of touchpoints.

Mobile Applications

While there are around 265 live mobile money deployments in the world—with another 102 planned[1]—only M-Pesa has served as a basis for the development of significant value-added services by app developers who enjoy the benefits of low-cost, extensive outreach and convenience to customers. In Sri Lanka, the MNOs are taking small steps in the right direction by enabling government pension disbursements, tax payments, and international remittances. In some cities, near-field communications (NFC)–enabled, contactless smart cards can be topped up and used to pay the bus fare.

Card-Based Grant Disbursement Systems

Digitizing payment of government grants, salary payments, and all government-to-persons (G2P) operations (including pensions) is one of the most effective ways of getting poor people into the financial space. Digitizing these payments can address corruption, ghost accounts, leakages, and political intervention, as well as enhancing financial discipline in the delivery of government payments.

Brazil and Mexico have digitized government payments, while South Africa has started the process with a biometric-enabled unique identification (ID) system.

However, with the exception of India's Aadhaar-enabled government benefit transfer program, all other South Asian countries use cash, commercial or state banks, and microfinance institutions for grant disbursement. Digitizing these payments would significantly enhance efficiency and reduce leakages, as well as enhance financial inclusion of the beneficiaries.

Customer Level

Unique Identification

A unique ID is a key element in digital finance. It enables basic KYC subscriber identification module (SIM) registration and avoids paper-based, tardy AML/CFT and KYC and CDD processes that tend to exclude and discourage the unbanked poor. The plethora of documents required at the time of opening mobile or e-money accounts has put off the poor and pushed them to over-the-counter (OTC) transactions to avoid paper-based procedures. Bangladesh's bKash system and Pakistan's Telenor systems are experiencing OTC-related issues.

In the South Asia region, except for Bangladesh and Nepal, other countries have unique ID. Pakistan has the National Database and Registration Authority (NADRA)-issued Computerized National Identity Card (CNIC), and Afghanistan is planning to roll out the electronic Tazkera. India has Aadhaar ID, which is biometric-enabled and linked to bank accounts. In addition, Sri Lanka mobile subscribers have a Digital Connect online ID that can be used for authentication for online purchases.

Note

1. Data from GSMA's Mobile Money Tracker: http://www.gsma.com/mobilefordevelopment/programmes/mobile-money-for-the-unbanked/insights/tracker.

Bibliography

di Castri, S. 2013. "What Could We Learn from Nigeria Barring MNOs from Participating in the Mobile Money Market?" Groupe Speciale Mobile Association blog, April 29. http://www.gsma.com/mobilefordevelopment/programme/mobile-money/what-could-we-learn-from-nigeria-barring-mnos-from-participating-in-the-mobile-money-market.

Klapper, L., and D. Singer. 2014. "The Opportunities of Digitizing Payments: How Digitization of Payments, Transfers, and Remittances Contributes to the G20 Goals of Broad-Based Economic Growth, Financial Inclusion, and Women's Economic Empowerment." Report No. 90305, prepared for the G20 Australian Presidency, World Bank, Washington, DC.

CHAPTER 9

Conclusions

Introduction

In the developing world as a whole, lack of money is the most commonly reported reason for being unbanked. Other reasons for not having access to formal financial services (through a bank, credit union, savings and credit cooperative, post office, or microfinance institution) include lack of proximity and the costs associated with maintaining an account. The wide accessibility of mobile-phone and digital technologies offers the potential to reduce disparities by gender and area (urban or rural).

Technology-based solutions offer a tremendous opportunity to transform the landscape of access to financial services for typically underserved groups, such as the poor, women, and remote populations. Digital or e-money can be enabled by mobile phones, branchless banking, point-of-sale (POS) transactions, prepaid or smart cards, and well-organized agent networks. These decentralized modalities have the capacity to reach the unbanked masses in a safe, simple, reliable, convenient, and cost-effective manner, enabling them to manage small transactions, including personal and government payments as well as remittance transfers.

Greater financial inclusion means higher potential for the poor to participate and share in their country's economic growth. Although the services available through e-money at present do not represent the full range of services available through financial institutions, they nevertheless enable those at the bottom of the pyramid to formally conduct certain basic financial transactions, and they can increasingly be linked to bank accounts and other means of accessing a more complete range of financial services.

This study has highlighted Kenya, South Africa, Sri Lanka, and Thailand as countries where the private sector and nonbank entities have applied technology and innovative thinking to successfully address inclusion issues, supported by flexibly designed policies and regulations.

Kenya more than doubled the rate of financial inclusion in five years to reach nearly 70 percent of the adult population as a direct result of innovations associated with the M-Pesa mobile money application, which has evolved into the country's dominant retail payment platform. M-Pesa has especially benefited the poor and unbanked, who previously had limited and costly access to traditional bank and financial infrastructure.

South Africa has issued some 10 million biometrically secure debit MasterCards as the platform for social transfer payments, thereby extending financial access to 16 million poor beneficiaries, with the country's banked and financially included population reaching 75 percent and 86 percent, respectively.

Sri Lanka's government and central bank proactively developed the country's legislative framework to support establishment of an excellent payment systems infrastructure, and possibly the best regulatory framework in the region to govern e-money for e-commerce and e-government. The result is the world's first end-to-end interoperable mobile payment solution, reaching over 83 percent of the population.

Thailand's efficient coordination of strategies and policies toward payment services and reduction of infrastructure costs—partly through the deployment of thousands of multicapacity automated teller machines (ATMs) and automated deposit machines (ADMs) throughout the country—has led to nearly 100 percent financial inclusion.

These findings, complemented by experiences from several other countries that are laying the foundations, reveal how e-money and other digital technologies can transform the financial inclusion landscape for the poor and underserved. This chapter summarizes the key lessons from the case study experiences that can be applied broadly to expand financial inclusion throughout the developing world.

Role of Governments and Regulators

While governments in principle have an interest in promoting financial inclusion as one instrument of poverty reduction and growth, their ability to take the lead may be constrained by lack of hard information on the results of new interventions, and their views may be influenced by powerful vested interests. Technologies that enable financial transactions to be conducted over mobile phones, POS, and other digital devices are disruptive to traditional financial systems because they bring in mobile network operators (MNOs) and other agents who are not licensed or regulated by the central bank authorities. Since the principal task of financial regulators is to minimize risks to the security and stability of the financial system, they are often understandably reluctant to introduce innovative technologies that might heighten such risks. Thus, the starting point is to document the roles that governments and regulators have played in promoting and facilitating the introduction and scaling-up of innovative digital technologies for expanding access to financial transactions, as well as cases in which lack of needed action has stalled development.

Three key roles for governments and regulators are to set the legal and regulatory ground rules; establish a unique national identification (ID) system;

and incentivize or mobilize participation in digital payments, especially by using social welfare grant programs to drive demand and use.

A Conducive Legal and Regulatory Environment

Kenya's M-Pesa mobile money application would not have occurred had the Central Bank of Kenya (CBK) not been willing to give a no-objection letter to an experimental (and at the time, unproven) mobile money scheme in the interests of financial inclusion—despite some resistance from commercial banks and politicians. CBK then drew on experience with the scheme to eventually develop an appropriate legal and regulatory framework, which facilitated the entry of other nonbank players and their agents.

In Sri Lanka, the government and the Central Bank of Sri Lanka (CBSL) adopted a different but also proactive approach by early on enacting laws and regulations that created a level playing field, where both banks and nonbanks (including MNOs) could offer digital money. CBSL established distinct guidelines for an MNO-led model, as well as a bank-led model, opening up the field to a broader and growing range of providers with different value propositions. Leveling the playing field has the advantage of enabling private sector players such as MNOs to offer competitive products and services efficiently and at a lower cost. This policy approach has facilitated the launch of the world's first end-to-end interoperable mobile payment solution.

The Philippines Ministry of Finance and Central Bank used a "test-learn-regulate" model—although success has been constrained by overly restrictive know-your-customer (KYC) registration requirements and lack of clarity on the regulatory mandate regarding e-money, as well as the absence of a unique ID system. The Bank of Thailand likewise is seeking an appropriate balance between extending outreach, managing risks to the system, and protecting consumers. It has agreed to extend its supervisory and regulatory mandate to specialized financial institutions that offer extensive coverage in rural and poor communities that lack a commercial bank branch.

The conclusion is that the government and central bank must at least play a supporting role in establishing conducive conditions—and sometimes can play a lead, game-changing role—but that different approaches may be appropriate in different situations. A flexible approach to innovation can allow experimentation with new technology, leading to well-adapted laws and regulations based on experience. On the other hand, establishing clear rules that level the playing field for both financial institutions and MNOs to apply digital technologies can foster both competition and cooperation among private players to achieve the best value proposition for both providers and consumers.

A Unique National ID System

India's unique ID program has enabled direct benefit transfer programs to be implemented through bank accounts. While not specifically linked to mobile technologies, this opens up the space for both MNOs and financial institutions to offer low-cost products to a huge, previously underserved market. In contrast, the Philippines' implementation difficulties were exacerbated by the lack of a national ID.

Incentivizing and Mobilizing Widespread Participation

India's Jan Dhan Yojana (JDY) program demonstrates the transformative power of a highly publicized campaign plus incentives for people to open up bank accounts. This built on the unique ID system, which was critical to overcoming KYC challenges. Although the potential of the scheme is yet to be fully realized, it has opened up a huge opportunity for India's low-income population to be at least nominally "banked," making this market attractive for private providers of both financial and telecommunications services.

The government of Thailand has taken advantage of widespread, publicly supported specialized financial institutions such as agricultural cooperatives and the Village Fund to extend microcredit to the underserved population, especially in rural areas.

The lesson is that, with sufficient political will, governments can undertake direct actions that help transform the financial inclusion landscape. These include a unique ID system; incentives and promotional efforts to get people to sign up for bank, e-money, or other accounts that give them access to financial products; and expanding the network of touchpoints accessible to the rural poor.

Digitizing Social Grant Programs: A Game Changer

Brazil's Bolsa Família program built on a merger of four cash grant schemes to become the world's largest conditional cash transfer program, reaching about 30 percent of the population. A single national registry system, managed by a state bank, regularly reviews eligibility, substituting for a unique ID. About 15 percent of beneficiaries obtain their grants through bank accounts that can be opened under simplified conditions and offer more flexibility than the "Social Cards."

South Africa's seven social security grant programs reach some 31 percent of the population. The South African Social Security Agency brought together a technology company, a bank, and MasterCard to develop a streamlined disbursement and authentication methodology using biometric chip technology embedded on a card that can operate offline as well as online, making it suitable for use in battery-operated smart card readers in rural areas. The card is "open" and also allows its users to do other financial transactions using ATMs and POS technology in retail outlets—thereby directly enhancing the financial access of its 16 million users beyond just receiving grants. The system also yielded significant efficiency gains and savings by eliminating duplicate and ghost beneficiaries. The security of biometric and password protection is cited by beneficiaries as a tremendous benefit that prevents theft and fraud.

Similar efforts in Mexico to deliver benefits through a "closed" debit card proved less successful in terms of increasing financial access because of the failure to authorize a large enough number of agencies for cash withdrawal to make it sufficiently convenient to the beneficiaries. Unlike in South Africa, the "closed-loop" Oportunidades card can only be used to receive a cash-out of the social grant benefit from a Bansefi Bank officer on a specified day and hour.

Because the bank accounts established in massive numbers under India's JDY program are linked to the unique Aadhaar ID, they now offer an efficient vehicle for various state and national welfare schemes to deliver subsidies directly and quickly to the authenticated beneficiaries. In addition, the program presents an important potential market for the services of private mobile money and financial service providers.

The conclusion is that expansion of "e-government" to reach the bottom of the pyramid by utilizing digital technologies such as mobile apps, prepaid cards, and direct transfers into bank accounts can be a particularly effective way to incentivize adoption of these digital means. Besides cost-effectively extending financial inclusion, these technologies can also dramatically increase the administrative efficiency of identifying and targeting those eligible to receive social grants, thereby reducing leakages during the disbursement process.

Coordinated Action, Common Platforms, and Interoperability

Although laws and regulations are part of the macro environment, they are implemented by agencies that can be considered as meso-level institutions. Part of Sri Lanka's success is attributable to close collaboration between CBSL and the Telecommunications Regulatory Commission of Sri Lanka. By coordinating proactively and converging the service delivery and customer protection paradigms of the telecommunications and banking sectors, the two regulatory authorities developed a comprehensive yet flexible mobile money regulatory and oversight framework. The National Payments Corporation of India has facilitated greater adoption of mobile banking by establishing a National Unified unstructured supplementary service data (USSD) Platform that allows every customer to access banking services with a single number across all banks—irrespective of the telecom service provider, mobile handset make, or the region.

Although interoperability is not a required precondition—Kenya's M-Pesa succeeded without it, as a closed-loop proprietary system—it can facilitate greater competition as a spur to outreach and price reduction. Indeed, Kenya recently introduced regulations requiring interoperability with other payment systems, both national and international, leveling the playing field for other providers. In Thailand, interoperability has been successfully achieved through the National Interbank Transaction Management and Exchange Company, which supports all types of electronic payments and funds transfer from various channels including ATMs, counters, Internet, phone, and mobile channels, utilizing a secure and very efficient open platform. This development was strongly supported by the Thai Bankers' Association, which has proven to be a valuable meso-level organization that partners with the Bank of Thailand to further financial inclusion. Pakistan likewise has taken a step toward interoperability by linking its Easypaisa mobile money service with the existing banking structure, enabling customers to transfer funds between a number of banks.

In conclusion, *how* regulations are implemented proves to be more important in facilitating or retarding development than how well they are drafted. E-money is unique in that it overlaps different regulatory domains, thus running the risk of a mismatch of regulations and how they are interpreted by different agencies. Hence it is essential to establish coordination mechanisms for the regulatory agencies governing both the financial and the telecommunications sectors. Also, establishing a payments platform that is common to all banks instead of each bank having to develop its own platform helps banks to focus on customers. Furthermore, measures to promote interoperability are potential game changers to hasten the spread of efficient, affordable mobile-phone and financial services to the previously underserved masses.

Outreach by Retail Institutions

In Sri Lanka, the MNOs responded positively and cooperatively to the favorable environment that had been established by the central bank, which enabled the MNO-led mobile money system to work with complete neutrality across MNO mobile money providers. In an unprecedented move, the leading MNO, Dialog Axiata, invited the other MNOs to share its eZ Cash platform and set up a system for separate mobile wallets to be managed and interoperated on a common platform. The result is the first mobile money system in the world to be end-to-end interoperable across multiple service providers. The trustee bank arrangement protects consumer funds by ring-fencing the trust account and holding it in receivership in case both the MNO and the custodian bank fail.

In Indonesia, too, the three large MNOs decided to work together and pool resources rather than try to compete with each other to cover the geographically challenging terrain. Customers of the three MNOs can transfer and cash out money from any location across each other's network. In this case, however, they decided to keep their own identities by way of a multi-wallet interoperable arrangement.

The success of M-Pesa was due in large part to Safaricom already having an extensive network of agents selling airtime and their effective management of the recruitment and training of agents needed to support rapid scaling-up. A different approach was taken by the government of India, which decided to set up special payment banks and small finance banks in unbanked and underbanked regions. It remains to be seen whether this approach can be sufficiently cost-effective to compete with other payment mechanisms and thereby draw more of the population into these banks.

A key conclusion is that increasing accessibility of mobile money and financial services depends on the efforts of the retail institutions—banks and other financial institutions as well as MNOs—to develop and deliver affordable services that meet the needs of the previously unserved or underserved population.

Experience shows that private providers *can* work cooperatively to make common elements of the system work cost-effectively while competing for customers on the retail end, when the infrastructure and incentives have been established to level the playing field and provide neutrality across providers. Hence it is incumbent on governments and central banks to lay the regulatory and infrastructural foundation for private actors to do so profitably and to facilitate market forces to drive the players toward the desired results.

Agents represent a further key element in utilizing e-money to extend the outreach of financial services, given that the number of mobile money agents is typically 4 to 10 times the number of bank branches. However, managing agent networks requires a delicate balance between ensuring the profitability of each agent and providing enough touchpoints that consumers can be served quickly and efficiently.

Increasing Accessibility for Customers

Thailand's use of retail stores has been a major factor in the rapid expansion of mobile money by bringing services close to the customer. For example, at every 7-Eleven store, which are widely distributed throughout remote rural areas as well as in the city, one can send and retrieve money, pay utility bills, purchase an airline ticket, or make a bank deposit. Kenya has had some success with adapting the mobile-phone platform to facilitate microfinance programs and various forms of insurance (life, accident, health, and crop), as well as pensions for informal workers. Mobile POS devices offer the potential to greatly expand access to financial services through branchless banking.

On the other hand, countries that have rigidly applied KYC and customer due diligence (CDD) requirements at the bottom of the pyramid have tended to have disappointing rates of enrollment. More successful have been countries such as Sri Lanka that have adopted a tiered approach, with much lighter requirements for accounts with restricted uses and low-value transactions. Sri Lanka has also dealt with concerns about the security of information by establishing the world's first multioperator Mobile Connect solution that provides an independent means for authenticating a user to any service provider without requiring a password.

An important conclusion from the customer point of view is that the direct cost of transactions is only one among several determinants of the value proposition of a mobile money or financial service. Proximity and convenience are often overriding concerns. Simplicity and ease of registration are also important. Although not yet a major concern to the low-end consumers, the security and possible misuse of the information provided represents an important risk to systems targeting millions of new customers—a risk that needs to be addressed in moving toward a cash-lite economic ecosystem. Overall, the linkage of mobile technologies with financial services offers many opportunities to improve the customer experience and provide more (or more efficient) services through new applications.

The Journey toward a Cash-Lite Society: Coordination and Balance

Two key steps toward a cash-lite, financially inclusive system are digitizing the payments and digitizing the money. The implied strategic question is whether to begin by democratizing bank accounts through massive enrollment, or by making e-money readily accessible over mobile-phone technology. In the countries studied, both bank-led and MNO-led models have been effective; either can create conditions that facilitate expansion of the other. Whether the potential is then realized depends heavily on other conditions, such as the existence of a unique national ID and the regulations governing who can do what and how banks and MNOs can link up (or compete).

Although governments and regulators clearly play a key role in the successful expansion of financial inclusion by leveraging digital technologies, again, no single strategic approach emerges: much depends on country circumstances. In Kenya, regulatory flexibility to permit experimentation with mobile-phone technology was the leading edge. In India, government incentives and the campaign for opening bank accounts is the leading game changer, with establishment of a unique national ID system as a fundamental precondition.

No matter how conducive the framework, the success of efforts to promote financial inclusion ultimately depends on the readiness of the previously unserved or underserved population to participate. Experiences in the countries studied show that cost and convenience are the most critical factors. The key challenge that digital technologies are helping to address is to make small transactions feasible at prices that are both affordable to the users and profitable to the suppliers. The equally critical factor of proximity to the client is being successfully addressed in some countries through low-cost options for the banking system to reach out, such as agent banking and POS devices, and in others by making financial "wallets" available through mobile phones, which already have high penetration. Other factors affecting the value proposition to consumers include the increased safety of going cashless, security of information, simplicity (including formats readily accessible to the nonliterate population), and reliability.

Where to begin or to target interventions depends very much on the country context and the existing state of laws and regulations governing mobile money and banking, the extent of penetration of bank accounts and mobile telephony, the existence of national ID, and cultural and other factors. It also depends on which technologies and innovations may be available at any given time to address the needs of the targeted population and the challenges facing implementation. So, each country that wishes to use the potential of e-money technologies to advance financial inclusion will need to undertake an assessment of its current policy, legal and regulatory framework, institutional environment, characteristics of targeted consumer groups, and the state of technology in the country relative to what is available internationally.

To move forward effectively, each country must also establish effective coordination and consultation mechanisms among the key players at all levels: macro (government policy makers, central banks, and other regulators); meso- and micro-level institutions (financial institutions and MNOs, and their associations and facilitating organizations); and consumers (through advocacy agencies and survey data). Key objectives for such coordination are to balance the interests of the different actors, ensure that the playing field is level, and, above all, strike the right balance among expanding outreach, mitigating risks to the stability and security of the system, and protecting the consumers.

The lessons provided in this study are intended to help flatten the learning curve for taking advantage of e-money and other innovations to extend access to financial services, especially for South Asian and other developing countries. By establishing a conducive legal and regulatory framework and providing appropriate incentives to both private providers and customers, countries can accelerate financial inclusion to promote a more inclusive society and shared prosperity.

APPENDIX A

Findex Data for Selected Countries

Table A.1 India against Benchmarks for South Asia and Lower-Middle-Income Countries
Percentage

Population, ages 15+ years: 887.9 million

GNI per capita: US$1,570

Survey item	*India*	*South Asia*	*Lower-middle-income*	*World average*
Account (ages 15+ years)				
All adults	53.1	46.4	42.7	61.5
Women	43.1	37.4	36.3	58.1
Adults belonging to the poorest 40%	43.9	38.1	33.2	54.0
Young adults (ages 15–24 years)	43.2	36.7	34.7	46.3
Adults living in rural areas	50.1	43.5	40.0	56.7
Financial institution account (ages 15+ years)				
All adults, 2014	52.8	45.5	41.8	60.7
All adults, 2011	35.2	32.3	28.7	50.6
Mobile account (ages 15+ years)				
All adults	2.4	2.6	2.5	2.0
Access to financial institution account (ages 15+ years)				
Has debit card, 2014	22.1	18.0	21.2	40.1
Has debit card, 2011	8.4	7.2	10.1	30.5
ATM is the main mode of withdrawal (% with an account), 2014	33.1	31.1	42.4	—
ATM is the main mode of withdrawal (% with an account), 2011	18.4	16.9	28.1	48.3
Use of account in the past year (ages 15+ years)				
Used an account to receive wages	4.0	3.5	5.6	17.7
Used an account to receive government transfers	3.6	3.1	3.3	8.2
Used a financial institution account to pay utility bills	3.4	2.7	3.1	16.7
Other digital payment in the past year (ages 15+ years)				
Used a debit card to make payments	10.7	8.5	9.6	23.2
Used a credit card to make payments	3.4	2.6	2.8	15.1
Used the Internet to pay bills or make purchases	1.2	1.2	2.6	16.6

table continues next page

Table A.1 India against Benchmarks for South Asia and Lower-Middle-Income Countries *(continued)*
Percentage

Population, ages 15+ years: 887.9 million *GNI per capita: US$1,570*				
Survey item	*India*	*South Asia*	*Lower-middle-income*	*World average*
Domestic remittances in the past year (ages 15+ years)				
Sent remittances	9.9	10.7	14.2	—
Sent remittances via a financial institution (% senders)	—	20.1	30.9	—
Sent remittances via a mobile phone (% senders)	—	7.7	7.7	—
Sent remittances via a money transfer operator (% senders)	—	13.7	18.3	—
Received remittances	9.8	12.2	17.8	—
Received remittances via a financial institution (% recipients)	—	15.8	26.0	—
Received remittances via a mobile phone (% recipients)	—	4.7	5.7	—
Received remittances via a money transfer operator (% recipients)	—	9.8	16.6	
Savings in the past year (ages 15+ years)				
Saved at a financial institution, 2014	14.4	12.7	14.8	27.4
Saved at a financial institution, 2011	11.6	11.1	11.1	22.6
Saved using a savings club or person outside the family	8.8	8.8	12.4	—
Saved any money	38.3	36.2	45.6	56.5
Saved for old age	9.9	9.1	12.6	23.9
Saved for a farm or business	7.0	7.3	11.8	13.8
Saved for education or school fees	16.0	14.6	20.0	22.3
Credit in the past year (ages 15+ years)				
Borrowed from a financial institution, 2014	6.4	6.4	7.5	10.7
Borrowed from a financial institution, 2011	7.7	8.7	7.3	9.1
Borrowed from family or friends	32.3	31.4	33.1	26.2
Borrowed from a private informal lender	12.6	10.9	8.5	4.6
Borrowed any money	46.3	46.7	47.4	42.4
Borrowed for a farm or business	9.0	8.6	9.2	7.1
Borrowed for education or school fees	9.7	8.9	10.1	7.7
Outstanding mortgage at a financial institution	3.7	3.8	4.7	10.4

Source: Global Findex 2014 Survey, http://datatopics.worldbank.org/financialinclusion/.
Note: ATM = automated teller machine; GNI = gross national income; — = not available.

Table A.2 Indonesia against Benchmarks for East Asia and Pacific and Lower-Middle-Income Countries
Percentage

Population, ages 15+ years: 177.7 million *GNI per capita: US$3,580*				
Survey item	*Indonesia*	*East Asia and Pacific*	*Lower-middle-income*	*World average*
Account (ages 15+ years)				
All adults	36.1	69.0	42.7	61.5
Women	37.5	67.0	36.3	58.1
Adults belonging to the poorest 40%	22.2	60.9	33.2	54.0
Young adults (ages 15–24 years)	35.2	60.7	34.7	46.3
Adults living in rural areas	28.7	64.5	40.0	56.7

table continues next page

Table A.2 Indonesia against Benchmarks for East Asia and Pacific and Lower-Middle-Income Countries *(continued)*
Percentage

Population, ages 15+ years: 177.7 million

GNI per capita: US$3,580

Survey item	*Indonesia*	*East Asia and Pacific*	*Lower-middle-income*	*World average*
Financial institution account (ages 15+ years)				
All adults, 2014	35.9	68.8	41.8	60.7
All adults, 2011	19.6	55.1	28.7	50.6
Mobile account (ages 15+ years)				
All adults	0.4	0.4	2.5	2.0
Access to financial institution account (ages 15+ years)				
Has debit card, 2014	25.9	42.9	21.2	40.1
Has debit card, 2011	10.5	34.7	10.1	30.5
ATM is the main mode of withdrawal (% with an account), 2014	70.9	53.3	42.4	—
ATM is the main mode of withdrawal (% with an account), 2011	51.1	37.0	28.1	48.3
Use of account in the past year (ages 15+ years)				
Used an account to receive wages	6.6	15.1	5.6	17.7
Used an account to receive government transfers	3.0	8.1	3.3	8.2
Used a financial institution account to pay utility bills	2.9	11.8	3.1	16.7
Other digital payment in the past year (ages 15+ years)				
Used a debit card to make payments	8.5	14.8	9.6	23.2
Used a credit card to make payments	1.1	10.8	2.8	15.1
Used the Internet to pay bills or make purchases	5.1	15.6	2.6	16.6
Domestic remittances in the past year (ages 15+ years)				
Sent remittances	17.9	16.6	14.2	—
Sent remittances via a financial institution (% senders)	52.4	36.9	30.9	—
Sent remittances via a mobile phone (% senders)	3.6	8.7	7.7	—
Sent remittances via a money transfer operator (% senders)	8.7	18.5	18.3	—
Received remittances	31.0	20.6	17.8	—
Received remittances via a financial institution (% recipients)	36.3	29.0	26.0	—
Received remittances via a mobile phone (% recipients)	0.2	4.9	5.7	—
Received remittances via a money transfer operator (% recipients)	7.9	15.8	16.6	—
Savings in the past year (ages 15+ years)				
Saved at a financial institution, 2014	26.6	36.5	14.8	27.4
Saved at a financial institution, 2011	15.3	28.5	11.1	22.6
Saved using a savings club or person outside the family	25.2	6.0	12.4	—
Saved any money	69.3	71.0	45.6	56.5
Saved for old age	27.1	36.5	12.6	23.9
Saved for a farm or business	22.6	21.3	11.8	13.8
Saved for education or school fees	33.3	30.7	20.0	22.3
Credit in the past year (ages 15+ years)				
Borrowed from a financial institution, 2014	13.1	11.0	7.5	10.7
Borrowed from a financial institution, 2011	8.5	8.6	7.3	9.1
Borrowed from family or friends	41.5	28.3	33.1	26.2
Borrowed from a private informal lender	2.9	2.5	8.5	4.6
Borrowed any money	56.6	41.2	47.4	42.4
Borrowed for a farm or business	11.7	8.3	9.2	7.1
Borrowed for education or school fees	12.2	7.1	10.1	7.7
Outstanding mortgage at a financial institution	5.5	8.0	4.7	10.4

Source: Global Findex 2014 Survey, http://datatopics.worldbank.org/financialinclusion/.
Note: ATM = automated teller machine; GNI = gross national income; — = not available.

Table A.3 Kenya against Benchmarks for Sub-Saharan Africa and Low-Income Countries
Percentage

Population, ages 15+ years: 25.6 million *GNI per capita: US$1,160*				
Survey item	*Kenya*	*Sub-Saharan Africa*	*Low-income*	*World average*
Account (ages 15+ years)				
All adults	74.7	34.2	27.5	61.5
Women	71.1	29.9	23.9	58.1
Adults belonging to the poorest 40%	63.4	24.6	19.4	54.0
Young adults (ages 15–24 years)	66.4	25.9	20.2	46.3
Adults living in rural areas	73.0	29.2	24.8	56.7
Financial institution account (ages 15+ years)				
All adults, 2014	55.2	28.9	22.3	60.7
All adults, 2011	42.3	23.9	21.1	50.6
Mobile account (ages 15+ years)				
All adults	58.4	11.5	10.0	2.0
Access to financial institution account (ages 15+ years)				
Has debit card, 2014	34.7	17.9	6.6	40.1
Has debit card, 2011	29.9	15.0	6.3	30.5
ATM is the main mode of withdrawal (% with an account), 2014	52.7	53.8	20.2	—
ATM is the main mode of withdrawal (% with an account), 2011	69.2	51.7	19.7	48.3
Use of account in the past year (ages 15+ years)				
Used an account to receive wages	18.0	7.3	3.2	17.7
Used an account to receive government transfers	6.4	3.8	1.0	8.2
Used a financial institution account to pay utility bills	5.8	2.8	0.9	16.7
Other digital payment in the past year (ages 15+ years)				
Used a debit card to make payments	11.2	8.7	2.1	23.2
Used a credit card to make payments	2.7	1.9	0.6	15.1
Used the Internet to pay bills or make purchases	4.7	2.4	1.2	16.6
Domestic remittances in the past year (ages 15+ years)				
Sent remittances	53.0	28.7	18.3	—
Sent remittances via a financial institution (% senders)	16.2	31.0	15.4	—
Sent remittances via a mobile phone (% senders)	92.0	30.8	42.8	—
Sent remittances via a money transfer operator (% senders)	8.9	21.0	14.1	—
Received remittances	61.0	37.2	25.6	—
Received remittances via a financial institution (% recipients)	14.2	26.6	13.0	—
Received remittances via a mobile phone (% recipients)	88.8	27.6	33.8	—
Received remittances via a money transfer operator (% recipients)	9.9	22.1	14.8	—
Savings in the past year (ages 15+ years)				
Saved at a financial institution, 2014	30.2	15.9	9.9	27.4
Saved at a financial institution, 2011	23.3	14.3	11.5	22.6
Saved using a savings club or person outside the family	39.9	23.9	16.3	—
Saved any money	76.1	59.6	46.5	56.5
Saved for old age	17.9	9.8	8.3	23.9
Saved for a farm or business	36.2	22.7	16.7	13.8
Saved for education or school fees	39.3	22.9	16.6	22.3

table continues next page

Table A.3 Kenya against Benchmarks for Sub-Saharan Africa and Low-Income Countries *(continued)*
Percentage

Population, ages 15+ years: 25.6 million				
GNI per capita: US$1,160				
Survey item	*Kenya*	*Sub-Saharan Africa*	*Low-income*	*World average*
Credit in the past year (ages 15+ years)				
Borrowed from a financial institution, 2014	14.9	6.3	8.6	10.7
Borrowed from a financial institution, 2011	9.7	4.8	11.7	9.1
Borrowed from family or friends	60.5	41.9	34.9	26.2
Borrowed from a private informal lender	7.3	4.7	6.5	4.6
Borrowed any money	79.2	54.5	52.5	42.4
Borrowed for a farm or business	24.3	12.8	12.2	7.1
Borrowed for education or school fees	33.5	12.3	10.9	7.7
Outstanding mortgage at a financial institution	12.1	5.2	4.1	10.4

Source: Global Findex 2014 Survey, http://datatopics.worldbank.org/financialinclusion/.
Note: ATM = automated teller machine; GNI = gross national income; — = not available.

Table A.4 The Philippines against Benchmarks for East Asia and Pacific and Lower-Middle-Income Countries
Percentage

Population, ages 15+ years: 64.8 million				
GNI per capita: US$3,270				
Survey item	*Philippines*	*East Asia and Pacific*	*Lower-middle-income*	*World average*
Account (ages 15+ years)				
All adults	31.3	69.0	42.7	61.5
Women	37.9	67.0	36.3	58.1
Adults belonging to the poorest 40%	17.8	60.9	33.2	54.0
Young adults (ages 15–24 years)	19.0	60.7	34.7	46.3
Adults living in rural areas	27.5	64.5	40.0	56.7
Financial institution account (ages 15+ years)				
All adults, 2014	28.1	68.8	41.8	60.7
All adults, 2011	26.6	55.1	28.7	50.6
Mobile account (ages 15+ years)				
All adults	4.2	0.4	2.5	2.0
Access to financial institution account (ages 15+ years)				
Has debit card, 2014	20.5	42.9	21.2	40.1
Has debit card, 2011	13.2	34.7	10.1	30.5
ATM is the main mode of withdrawal (% with an account), 2014	67.1	53.3	42.4	—
ATM is the main mode of withdrawal (% with an account), 2011	62.5	37.0	28.1	48.3
Use of account in the past year (ages 15+ years)				
Used an account to receive wages	6.3	15.1	5.6	17.7
Used an account to receive government transfers	4.0	8.1	3.3	8.2
Used a financial institution account to pay utility bills	1.0	11.8	3.1	16.7
Other digital payment in the past year (ages 15+ years)				
Used a debit card to make payments	11.9	14.8	9.6	23.2
Used a credit card to make payments	2.2	10.8	2.8	15.1
Used the Internet to pay bills or make purchases	3.5	15.6	2.6	16.6

table continues next page

Table A.4 The Philippines against Benchmarks for East Asia and Pacific and Lower-Middle-Income Countries *(continued)*
Percentage

Population, ages 15+ years: 64.8 million

GNI per capita: US$3,270

Survey item	*Philippines*	*East Asia and Pacific*	*Lower-middle-income*	*World average*
Domestic remittances in the past year (ages 15+ years)				
Sent remittances	21.3	16.6	14.2	—
Sent remittances via a financial institution (% senders)	17.0	36.9	30.9	—
Sent remittances via a mobile phone (% senders)	16.2	8.7	7.7	—
Sent remittances via a money transfer operator (% senders)	70.5	18.5	18.3	—
Received remittances	34.1	20.6	17.8	—
Received remittances via a financial institution (% recipients)	12.1	29.0	26.0	—
Received remittances via a mobile phone (% recipients)	10.8	4.9	5.7	—
Received remittances via a money transfer operator (% recipients)	58.0	15.8	16.6	—
Savings in the past year (ages 15+ years)				
Saved at a financial institution, 2014	14.8	36.5	14.8	27.4
Saved at a financial institution, 2011	14.7	28.5	11.1	22.6
Saved using a savings club or person outside the family	9.3	6.0	12.4	—
Saved any money	67.3	71.0	45.6	56.5
Saved for old age	24.5	36.5	12.6	23.9
Saved for a farm or business	22.9	21.3	11.8	13.8
Saved for education or school fees	41.9	30.7	20.0	22.3
Credit in the past year (ages 15+ years)				
Borrowed from a financial institution, 2014	11.8	11.0	7.5	10.7
Borrowed from a financial institution, 2011	10.5	8.6	7.3	9.1
Borrowed from family or friends	48.7	28.3	33.1	26.2
Borrowed from a private informal lender	13.5	2.5	8.5	4.6
Borrowed any money	69.7	41.2	47.4	42.4
Borrowed for a farm or business	13.6	8.3	9.2	7.1
Borrowed for education or school fees	29.9	7.1	10.1	7.7
Outstanding mortgage at a financial institution	4.9	8.0	4.7	10.4

Source: Global Findex 2014 Survey, http://datatopics.worldbank.org/financialinclusion/.
Note: ATM = automated teller machine; GNI = gross national income; — = not available.

Table A.5 South Africa against Benchmarks for Sub-Saharan Africa and Upper-Middle-Income Countries
Percentage

Population, ages 15+ years: 37.5 million

GNI per capita: US$7,410

Survey item	*South Africa*	*Sub-Saharan Africa*	*Upper-middle-income*	*World average*
Account (ages 15+ years)				
All adults	70.3	34.2	70.5	61.5
Women	70.4	29.9	67.3	58.1
Adults belonging to the poorest 40%	57.8	24.6	62.7	54.0
Young adults (ages 15–24 years)	53.5	25.9	58.1	46.3
Adults living in rural areas	70.0	29.2	68.8	56.7

table continues next page

Table A.5 South Africa against Benchmarks for Sub-Saharan Africa and Upper-Middle-Income Countries *(continued)*
Percentage

Population, ages 15+ years: 37.5 million

GNI per capita: US$7,410

Survey item	*South Africa*	*Sub-Saharan Africa*	*Upper-middle-income*	*World average*
Financial institution account (ages 15+ years)				
All adults, 2014	68.8	28.9	70.4	60.7
All adults, 2011	53.6	23.9	57.4	50.6
Mobile account (age 15+ years)				
All adults	14.4	11.5	0.7	2.0
Access to financial institution account (ages 15+ years)				
Has debit card, 2014	54.9	17.9	45.9	40.1
Has debit card, 2011	45.3	15.0	38.5	30.5
ATM is the main mode of withdrawal (% with an account), 2014	81.8	53.8	55.7	—
ATM is the main mode of withdrawal (% with an account), 2011	88.9	51.7	42.8	48.3
Use of account in the past year (ages 15+ years)				
Used an account to receive wages	26.8	7.3	18.1	17.7
Used an account to receive government transfers	28.2	3.8	9.6	8.2
Used a financial institution account to pay utility bills	12.2	2.8	12.3	16.7
Other digital payment in the past year (ages 15+ years)				
Used a debit card to make payments	40.8	8.7	19.9	23.2
Used a credit card to make payments	10.8	1.9	14.4	15.1
Used the Internet to pay bills or make purchases	7.6	2.4	15.3	16.6
Domestic remittances in the past year (ages 15+ years)				
Sent remittances	41.5	28.7	15.4	—
Sent remittances via a financial institution (% senders)	63.0	31.0	37.2	—
Sent remittances via a mobile phone (% senders)	17.6	30.8	8.8	—
Sent remittances via a money transfer operator (% senders)	56.6	21.0	19.7	—
Received remittances	54.2	37.2	7.8	—
Received remittances via a financial institution (% recipients)	54.9	26.6	29.8	—
Received remittances via a mobile phone (% recipients)	16.0	27.6	5.6	—
Received remittances via a money transfer operator (% recipients)	61.3	22.1	17.9	—
Savings in the past year (ages 15+ years)				
Saved at a financial institution, 2014	32.7	15.9	32.2	27.4
Saved at a financial institution, 2011	22.1	14.3	25.1	22.6
Saved using a savings club or person outside the family	30.6	23.9	4.9	—
Saved any money	66.4	59.6	62.7	56.5
Saved for old age	15.9	9.8	30.6	23.9
Saved for a farm or business	11.0	22.7	17.6	13.8
Saved for education or school fees	23.8	22.9	25.4	22.3
Credit in the past year (ages 15+ years)				
Borrowed from a financial institution, 2014	12.1	6.3	10.4	10.7
Borrowed from a financial institution, 2011	8.9	4.8	7.9	9.1
Borrowed from family or friends	71.2	41.9	24.0	26.2
Borrowed from a private informal lender	18.4	4.7	2.6	4.6
Borrowed any money	85.6	54.5	37.7	42.4
Borrowed for a farm or business	7.5	12.8	6.6	7.1
Borrowed for education or school fees	18.0	12.3	6.1	7.7
Outstanding mortgage at a financial institution	9.2	5.2	9.1	10.4

Source: Global Findex 2014 Survey, http://datatopics.worldbank.org/financialinclusion/.
Note: ATM = automated teller machine; GNI = gross national income; — = not available.

Table A.6 Sri Lanka against Benchmarks for South Asia and Lower-Middle-Income Countries
Percentage

Population, ages 15+ years: 15.3 million				
GNI per capita: US$3,170				
Survey item	*Sri Lanka*	*South Asia*	*Lower-middle-income*	*World average*
Account (ages 15+ years)				
All adults	82.7	46.4	42.7	61.5
Women	83.1	37.4	36.3	58.1
Adults belonging to the poorest 40%	79.8	38.1	33.2	54.0
Young adults (ages 15–24 years)	85.2	36.7	34.7	46.3
Adults living in rural areas	83.4	43.5	40.0	56.7
Financial institution account (ages 15+ years)				
All adults, 2014	82.7	45.5	41.8	60.7
All adults, 2011	68.5	32.3	28.7	50.6
Mobile account (ages 15+ years)				
All adults	0.1	2.6	2.5	2.0
Access to financial institution account (ages 15+ years)				
Has debit card, 2014	24.9	18.0	21.2	40.1
Has debit card, 2011	10.0	7.2	10.1	30.5
ATM is the main mode of withdrawal (% with an account), 2014	24.3	31.1	42.4	—
ATM is the main mode of withdrawal (% with an account), 2011	15.4	16.9	28.1	48.3
Use of account in the past year (ages 15+ years)				
Used an account to receive wages	7.1	3.5	5.6	17.7
Used an account to receive government transfers	5.3	3.1	3.3	8.2
Used a financial institution account to pay utility bills	1.1	2.7	3.1	16.7
Other digital payment in the past year (ages 15+ years)				
Used a debit card to make payments	10.4	8.5	9.6	23.2
Used a credit card to make payments	2.8	2.6	2.8	15.1
Used the Internet to pay bills or make purchases	1.6	1.2	2.6	16.6
Domestic remittances in the past year (ages 15+ years)				
Sent remittances	10.2	10.7	14.2	—
Sent remittances via a financial institution (% senders)	28.8	20.1	30.9	—
Sent remittances via a mobile phone (% senders)	0.0	7.7	7.7	—
Sent remittances via a money transfer operator (% senders)	3.5	13.7	18.3	—
Received remittances	16.2	12.2	17.8	—
Received remittances via a financial institution (% recipients)	30.2	15.8	26.0	—
Received remittances via a mobile phone (% recipients)	0.0	4.7	5.7	—
Received remittances via a money transfer operator (% recipients)	0.5	9.8	16.6	—
Savings in the past year (ages 15+ years)				
Saved at a financial institution, 2014	30.9	12.7	14.8	27.4
Saved at a financial institution, 2011	28.1	11.1	11.1	22.6
Saved using a savings club or person outside the family	10.4	8.8	12.4	—
Saved any money	45.2	36.2	45.6	56.5
Saved for old age	13.8	9.1	12.6	23.9
Saved for a farm or business	7.2	7.3	11.8	13.8
Saved for education or school fees	12.6	14.6	20.0	22.3

table continues next page

Table A.6 Sri Lanka against Benchmarks for South Asia and Lower-Middle-Income Countries *(continued)*
Percentage

Population, ages 15+ years: 15.3 million *GNI per capita: US$3,170*				
Survey item	*Sri Lanka*	*South Asia*	*Lower-middle-income*	*World average*
Credit in the past year (ages 15+ years)				
Borrowed from a financial institution, 2014	17.9	6.4	7.5	10.7
Borrowed from a financial institution, 2011	17.7	8.7	7.3	9.1
Borrowed from family or friends	9.0	31.4	33.1	26.2
Borrowed from a private informal lender	2.4	10.9	8.5	4.6
Borrowed any money	29.1	46.7	47.4	42.4
Borrowed for a farm or business	3.1	8.6	9.2	7.1
Borrowed for education or school fees	4.4	8.9	10.1	7.7
Outstanding mortgage at a financial institution	7.7	3.8	4.7	10.4

Source: Global Findex 2014 Survey, http://datatopics.worldbank.org/financialinclusion/.
Note: ATM = automated teller machine; GNI = gross national income; — = not available.

Table A.7 Thailand against Benchmarks for East Asia and Pacific and Upper-Middle-Income Countries
Percentage

Population, ages 15+ years: 54.8 million *GNI per capita: US$5,340*				
Survey item	*Thailand*	*East Asia and Pacific*	*Upper-middle-income*	*World average*
Account (ages 15+ years)				
All adults	78.1	69.0	70.5	61.5
Women	75.4	67.0	67.3	58.1
Adults belonging to the poorest 40%	72.0	60.9	62.7	54.0
Young adults (ages 15–24 years)	70.6	60.7	58.1	46.3
Adults living in rural areas	78.2	64.5	68.8	56.7
Financial institution account (ages 15+ years)				
All adults, 2014	78.1	68.8	70.4	60.7
All adults, 2011	72.7	55.1	57.4	50.6
Mobile account (ages 15+ years)				
All adults	1.3	0.4	0.7	2.0
Access to financial institution account (ages 15+ years)				
Has debit card, 2014	54.8	42.9	45.9	40.1
Has debit card, 2011	43.1	34.7	38.5	30.5
ATM is the main mode of withdrawal (% with an account), 2014	62.3	53.3	55.7	—
ATM is the main mode of withdrawal (% with an account), 2011	59.3	37.0	42.8	48.3
Use of account in the past year (ages 15+ years)				
Used an account to receive wages	8.3	15.1	18.1	17.7
Used an account to receive government transfers	9.0	8.1	9.6	8.2
Used a financial institution account to pay utility bills	1.7	11.8	12.3	16.7

table continues next page

Table A.7 Thailand against Benchmarks for East Asia and Pacific and Upper-Middle-Income Countries *(continued)*
Percentage

Population, ages 15+ years: 54.8 million

GNI per capita: US$5,340

Survey item	*Thailand*	*East Asia and Pacific*	*Upper-middle-income*	*World average*
Other digital payment in the past year (ages 15+ years)				
Used a debit card to make payments	7.9	14.8	19.9	23.2
Used a credit card to make payments	3.7	10.8	14.4	15.1
Used the Internet to pay bills or make purchases	4.4	15.6	15.3	16.6
Domestic remittances in the past year (ages 15+ years)				
Sent remittances	36.7	16.6	15.4	—
Sent remittances via a financial institution (% senders)	35.6	36.9	37.2	—
Sent remittances via a mobile phone (% senders)	2.0	8.7	8.8	—
Sent remittances via a money transfer operator (% senders)	25.3	18.5	19.7	—
Received remittances	46.4	20.6	17.8	—
Received remittances via a financial institution (% recipients)	28.6	29.0	29.8	—
Received remittances via a mobile phone (% recipients)	1.2	4.9	5.6	—
Received remittances via a money transfer operator (% recipients)	19.9	15.8	17.9	—
Savings in the past year (ages 15+ years)				
Saved at a financial institution, 2014	40.6	36.5	32.2	27.4
Saved at a financial institution, 2011	42.8	28.5	25.1	22.6
Saved using a savings club or person outside the family	8.4	6.0	4.9	—
Saved any money	80.5	71.0	62.7	56.5
Saved for old age	59.2	36.5	30.6	23.9
Saved for a farm or business	16.4	21.3	17.6	13.8
Saved for education or school fees	24.1	30.7	25.4	22.3
Credit in the past year (ages 15+ years)				
Borrowed from a financial institution, 2014	15.4	11.0	10.4	10.7
Borrowed from a financial institution, 2011	19.4	8.6	7.9	9.1
Borrowed from family or friends	31.1	28.3	24.0	26.2
Borrowed from a private informal lender	9.1	2.5	2.6	4.6
Borrowed any money	50.3	41.2	37.7	42.4
Borrowed for a farm or business	12.8	8.3	6.6	7.1
Borrowed for education or school fees	7.6	7.1	6.1	7.7
Outstanding mortgage at a financial institution	10.9	8.0	9.1	10.4

Source: Global Findex 2014 Survey, http://datatopics.worldbank.org/financialinclusion/.
Note: ATM = automated teller machine; GNI = gross national income; — = not available.

Environmental Benefits Statement

The World Bank Group is committed to reducing its environmental footprint. In support of this commitment, we leverage electronic publishing options and print-on-demand technology, which is located in regional hubs worldwide. Together, these initiatives enable print runs to be lowered and shipping distances decreased, resulting in reduced paper consumption, chemical use, greenhouse gas emissions, and waste.

We follow the recommended standards for paper use set by the Green Press Initiative. The majority of our books are printed on Forest Stewardship Council (FSC)–certified paper, with nearly all containing 50–100 percent recycled content. The recycled fiber in our book paper is either unbleached or bleached using totally chlorine-free (TCF), processed chlorine–free (PCF), or enhanced elemental chlorine–free (EECF) processes.

More information about the Bank's environmental philosophy can be found at http://www.worldbank.org/corporateresponsibility.

green press INITIATIVE

www.ingramcontent.com/pod-product-compliance
Lightning Source LLC
Chambersburg PA
CBHW080529220326
41599CB00032B/6252